Video Coding for Mobile Communications

Efficiency, Complexity, and Resilience

Signal Processing and its Applications

Video Coding for Mobile Communications
Efficiency, Complexity, and Resilience

Mohammed Ebrahim Al-Mualla
Etisalat College of Engineering,
Emirates Telecommunications Corporation (ETISALAT), U.A.E.

C. Nishan Canagarajah and David R. Bull
Image Communications Group,
Center for Communications Research,
University of Bristol, U.K.

ACADEMIC PRESS
An Imprint of Elsevier Science

Amsterdam   Boston   London   New York   Oxford   Paris   San Diego
San Francisco   Singapore   Sydney   Tokyo

Academic Press
An Imprint of Elsevier Science
525 B Street, Suite 1900, San Diego, California 92101-4495, USA
http://www.academicpress.com

Academic Press
An Imprint of Elsevier Science
Harcourt Place, 32 Jamestown Road, London NW1 7BY, UK
http://www.academicpress.com

Library of Congress Catalog Card Number: 2001098017

International Standard Book Number: 0-12-053079-1

PRINTED IN THE UNITED STATES OF AMERICA

02 03 04 05 06 EB 9 8 7 6 5 4 3 2 1

To my parents, brothers, and sisters
Mohammed E. Al-Mualla

To my family
C. Nishan Canagarajah

To Janice
David R. Bull

Contents

Preface

Scope and Purpose of the Book

Motivated by the vision of being able to communicate from anywhere at any time with any type of information, a natural convergence of mobile and multimedia is under way. This new area, called *mobile multimedia communications*, is expected to achieve unprecedented growth and worldwide commercial success.

Current second-generation mobile communication systems support a number of basic multimedia communication services. However, many technologically demanding problems need to be solved before *real-time mobile video communications* can be achieved. When such challenges are resolved, a wealth of advanced services and applications will be available to the mobile user. This book concentrates on three main challenges:

1. Higher coding efficiency

2. Reduced computational complexity

3. Improved error resilience

Mobile video communications is an interdisciplinary subject. Complete systems are likely to draw together solutions from different areas, such as video source coding, channel coding, network design, and semiconductor design, among others. This book concentrates on solutions based on *video source coding*. In this context, the book adopts a *motion-based approach*, where advanced motion estimation techniques, reduced-complexity motion estimation techniques, and motion-compensated error concealment techniques are used as possible solutions to the three challenges, respectively.

The idea of this book originated in 1997, when the first author was in the early stages of his Ph.D. studies. As a newcomer to the field, he started consulting a number of books to introduce himself to the fundamentals and standards of video source coding. He realized, however, that, for a beginner, most of these books seemed too long and too theoretical, with no treatment of some important practical and implementation issues. As he progressed further in his studies, the first author also realized that the areas of coding efficiency, computational complexity, and error resilience are usually treated separately. Thus, he always wished there was a book that provided a quick, easy, and practical introduction to the fundamentals and standards of video source coding and that brought together the areas of coding efficiency, computational complexity, and error resilience in a single volume. This is exactly the purpose of this book.

Structure of the Book

The book consists of 10 chapters. Chapter 1 gives a brief introduction to mobile video communications. It starts by discussing the main motivations and applications of mobile video communications. It then briefly introduces the challenges of higher coding efficiency, reduced computational complexity, and error resilience. The chapter then discusses some possible motion-based solutions. The remaining chapters of the book are organized into four parts. The first part introduces the reader to video coding, whereas the remaining three parts are devoted to the three challenges of coding efficiency, computational complexity, and error resilience.

Part I gives an introduction to video coding. It contains two chapters. Chapter 2 introduces some of the fundamentals of video source coding. It starts by giving some basic definitions and then covers both analog, and digital video along with some basic video coding techniques. It also presents the performance measures and the test sequences that will be used throughout the book. It then reviews both intraframe and interframe video coding methods.

Chapter 3 provides a brief introduction to video coding standards. Particular emphasis is given to the most recent standards, such as H.263 (and its extensions H.263+ and H.263++) and MPEG-4.

Part II concentrates on coding efficiency. It contains three chapters. Chapter 4 covers some basic motion estimation methods. It starts by introducing some of the fundamentals of motion estimation. It then reviews some basic motion estimation methods, with particular emphasis on the widely used block-matching methods. The chapter then presents the results of a comparative study between the different methods. The chapter also investigates the efficiency of motion estimation at very low bit rates, typical of mobile video communications. The aim is to decide if the added complexity of this process is justifiable, in terms of an improved coding efficiency, at such bit rates.

Chapter 5 investigates the performance of the more advanced warping-based motion estimation methods. The chapter starts by describing a general warping-based motion estimation method. It then considers some important parameters, such as the shape of the patches, the spatial transformation used, and the node tracking algorithm. The chapter then assesses the suitability of warping-based methods for mobile video communications. In particular, the chapter compares the efficiency and complexity of such methods to those of block-matching methods.

Chapter 6 investigates the performance of another advanced motion-estimation method, called multiple-reference motion-compensated prediction. The chapter starts by briefly reviewing multiple-reference motion estimation methods. It then concentrates on the long-term memory motion-compensated prediction technique. The chapter investigates the prediction gains and the coding efficiency of this technique at very low bit rates. The primary aim is to decide if the added complexity, increased motion overhead, and increased memory requirements of this technique are justifiable at such bit rates. The chapter also investigates the properties of multiple-reference block-motion fields and compares them to those of single-reference fields.

Part III of the book considers the challenge of reduced computational complexity. It contains two chapters. Chapter 7 reviews reduced-complexity motion estimation techniques. The chapter uses implementation examples and profiling results to highlight the need for reduced-complexity motion estimation. It then reviews some of the main reduced-complexity block-matching motion estimation techniques. The chapter then presents the results of a study comparing the different techniques.

Chapter 8 gives an example of the development of a novel reduced-complexity motion estimation technique. The technique is called the *simplex minimization search*. The development process is described in detail, and the technique is then tested within an isolated test environment, a block-based H.263-like codec, and an object-based MPEG-4 codec. In an attempt to reduce the complexity of multiple-reference motion estimation (investigated in Chapter 6), the chapter extends the simplex minimization search technique to the multiple-reference case. The chapter presents three different extensions (or algorithms) representing different degrees of compromise between prediction quality and computational complexity.

Part IV concentrates on error resilience. It contains two chapters. Chapter 9 reviews error resilience video coding techniques. The chapter considers the types of errors that can affect a video bitstream and examines their impact on decoded video. It then describes a number of error detection and error control techniques. Particular emphasis is given to standard error-resilience techniques included in the recent H.263+, H.263++, and MPEG-4 standards.

Chapter 10 gives examples of the development of error-resilience techniques. The chapter presents two temporal error concealment techniques. The first technique is based on motion field interpolation, whereas the second technique uses multihypothesis motion compensation to combine motion field interpolation with a boundary-matching technique. The techniques are then tested within both an isolated test environment and an H.263 codec. The chapter also investigates the performance of different temporal error concealment techniques when incorporated within a multiple-reference video codec. In particular, the chapter finds a combination of techniques that best recovers the spatial-temporal components of a damaged multiple-reference motion vector. In addition, the chapter develops a multihypothesis temporal concealment technique to be used with multiple-reference systems.

Audience for the Book

In recent years, mobile video communications has become an active and important research and development topic in both industry and academia. It is, therefore, hoped that this book will appeal to a broad audience, including students, instructors, researchers, engineers, and managers.

Chapter 1 can serve as a quick introduction for managers. Chapters 2 and 3 can be used in an introductory course on the fundamentals and the standards of video coding. The two chapters can also be used as a quick introduction for researchers and engineers working on video coding for the first time. More advanced courses on video coding can also utilize Chapters 4, 7, and 9 to introduce the students to issues in coding efficiency, computational complexity, and error resilience. The three chapters can also be used by researchers and engineers as an introduction and a guide to the relevant literature in the respective areas. Researchers and engineers will also find Chapters 5, 6, 8, and 10 useful as examples of the design, implementation, and testing of novel video coding techniques.

Acknowledgments

We are greatly indebted to past and present members of the Image Communications Group in the Center for Communications Research, University of Bristol, for creating an environment from which a book such as this could emerge. In particular, we would like to thank Dr. Przemysław Czerepiński for his generous help in all aspects of video research, Dr. Greg Cain for interesting discussions on implementation and complexity issues of motion estimation, Mr. Chiew Tuan Kiang for fruitful discussions on the efficiency of motion estimation at very low bit rates, and Mr. Oliver Sohm for providing the MPEG-4 results of Chapter 8.

We also owe a debt to Joel ClayPool and Angela Dooley from Academic Press who have shown a great deal of patience with us while we pulled this project together.

Mohammed Al-Mualla is deeply appreciative of the Emirates Telecommunications Corporation (Etisalat), United Arab Emirates, for providing financial support. In particular, he would like to thank Mr. Ali Al-Owais, CEO and President of Etisalat, and Mr. Salim Al-Owais, Manager of Etisalat College of Engineering.

He would also like to give his special thanks to his friends: Saif Bin Haider, Khalid Almidfa, Humaid Gharib, Abdul-Rahim Al-Maimani, Dr. Naim Dahnoun, Dr. M. F. Tariq, and Doreen Castañeda.

He would also like to give his most heartfelt thanks to his family and especially to his parents and his dear brother, Majid, for their unwavering love, support, and encouragement.

Mohammed Ebrahim Al-Mualla
Umm Al-Quwain, U.A.E.

C. Nishan Canagarajah,
David R. Bull
Bristol, U.K.

About the Authors

Mohammed E. Al-Mualla received the B. Eng. (Honors) degree in communications from Etisalat College of Engineering, Sharjah, United Arab Emirates (U.A.E.), in 1995, and the M.Sc. degree in communications and signal processing from the University of Bristol, U.K., in 1997. In 2000, he received the Ph.D. degree from the University of Bristol with a thesis titled "Video Coding for Mobile Communications: A Motion-Based Approach." Dr. Al-Mualla is currently an Assistant Professor at Etisalat College of Engineering. His research is focused on mobile video communications, with a particular emphasis on the problems of higher coding efficiency, reduced computational complexity, and improved error resilience.

C. Nishan Canagarajah received the B.A. (Honors) degree and the Ph.D. degree in digital signal processing techniques for speech enhancement, both from the University of Cambridge, Cambridge, U.K. He is currently a reader in signal processing at the University of Bristol, Bristol, U.K. He is also a coeditor of the book *Insights into Mobile Multimedia Communications* in the Signal Processing and Its Applications series from Academic Press. His research interests include image and video coding, nonlinear filtering techniques, and the application of signal processing to audio and medical electronics. Dr. Canagarajah is a committee member of the IEE Professional Group E5, and an Associate Editor of the IEE Electronics and Communication Journal.

David R. Bull is currently a professor of signal processing at the University of Bristol, Bristol, U.K., where he is the head of the Electrical and Electronics Engineering Department. He also leads the Image Communications Group in the Center for Communications Research, University of Bristol. Professor Bull has worked widely in the fields of 1- and 2-D signal processing. He has published over 150 papers and is a coeditor of the book *Insights into Mobile Multimedia Communications* in the Signal Processing and Its Applications series from Academic Press. Previously, he was an electronic systems engineer at Rolls Royce and a lecturer at the University of Wales, Cardiff, U.K. His recent research has focused on the problems of image and video communications, in particular, error-resilient source coding, linear and nonlinear filterbanks, scalable methods, content-based coding, and architectural optimization. Professor Bull is a member of the EPSRC Communications College and the Program Management Committee for the DTI/EPSRC LINK program in broadcast technology. Additionally, he is a member of the U.K. Foresight ITEC panel.

List of Acronyms

525/60	a television system with 525 vertical lines and a refresh rate of 60 Hz
625/50	a television system with 625 vertical lines and a refresh rate of 50 Hz
ACK	acknowledgment
ARQ	automatic repeat request
ASO	arbitrary slice ordering
ATM	asynchronous transfer mode
AV	average motion vector
AVO	audio visual object
BAB	binary alpha block
BCH	Bose-Chaudhuri-Hocquenghem
BDM	block distortion measure
BER	bit error rate
BM	boundary matching
BMA	block-matching algorithm
BMME	block-matching motion estimation
CAE	context-based arithmetic encoding
CCIR	international consultative committee for radio
CCITT	international telegraph and telephone consultative committee
CD	compact disc; conjugate directions

CDS	conjugate-directions search
CGI	control grid interpolation
CIF	common intermediate format
CMY	cyan, magenta, and yellow
CR	conditional replenishment
CSA	cross-search algorithm
DCT	discrete cosine transform
DF	displaced frame
DFA	differential frame algorithm
DFD	displaced-frame difference
DFT	discrete Fourier transform
DM	delta modulation
DMS	discrete memoryless source
DPCM	differential pulse code modulation
DS	dynamic sprites; diamond search
DSCQS	double stimulus continuous quality scale
DSIS	double stimulus impairment scale
DSP	digital signal processor
DWT	discrete wavelet transform
ECVQ	entropy-constrained vector quantization
EDGE	enhanced data rates for GSM evolution
EREC	error-resilience entropy code
ERPS	enhanced reference picture selection
EZW	embedded zero-tree wavelet
FD	frame difference
FEC	forward error correction
FFT	fast Fourier transform
FLC	fixed-length coding

FS	full search
FT	Fourier transform
GA	genetic algorithm
GMC	global motion compensation
GMS	genetic motion search
GOB	group of blocks
GOV	group of video object planes
GPRS	general packet radio service
GSM	group spécial mobile or global system for mobile
HD	horizontal difference
HDTV	high-definition television
HEC	header extension code
HMA	hexagonal matching algorithm
HVS	human visual system
IDCT	inverse discrete cosine transform
IEC	international electrotechnical commission
ISD	independent segment decoding
ISDN	integrated services digital network
ISO	international organization for standardization
ITU-R	international telecommunications union — radio sector
ITU-T	international telecommunications union — telecommunication standardization sector
JTC	joint technical committee
KLT	Krahunen-Loève transform
LBG	Linde-Buzo-Gray
LMS	least mean square
LOT	lapped orthogonal transform
LPE	low-pass extrapolation
LTM-MCP	long-term memory motion-compensated prediction

MAE	mean absolute error
MAP	maximum *a posteriori* probability
MB	macroblock
MC	motion compensation
MCP	motion-compensated prediction
ME	motion estimation
MFI	motion field interpolation
MHMC	multihypothesis motion compensation
MIPS	million instructions per second
ML	maximum likelihood
MPEG	moving picture experts group
MRF	Markov random field
MR-MCP	multiple-reference motion-compensated prediction
MSE	mean squared error
NACK	negative acknowledgment
NCCF	normalized cross-correlation function
NSS	n-steps search
NTSC	national television system committee
OMC	overlapped motion compensation
OTS	one-at-a-time search
PAL	phase alternation line
PCA	phase correlation algorithm
PDC	pel difference classification
PDE	partial distortion elimination
PRA	pel-recursive algorithm
PSNR	peak signal-to-noise ratio
PSTN	public switched telephone network

QAM	quadrature amplitude modulation
QCIF	quarter common intermediate format
QMF	quadrature mirror filter
QP	quantization parameter
QSIF	quarter source input format
RBMAD	reduced-bits mean absolute difference
R-D	rate-distortion
RGB	red, green, and blue
RLE	run-length encoding
RPS	reference picture selection
RS	rectangular slice
RVLC	reversible variable-length coding
SAD	sum of absolute differences
SC	subcommittee
SEA	successive elimination algorithm
SECAM	sequential couleur avec memoire
SG	study group
SIF	source input format
SM	simplex minimization
SMD	side-match distortion
SMS	simplex minimization search
SNR	signal-to-noise ratio
SPIHT	set partitioning in hierarchical trees
SQCIF	sub-QCIF
SSCQS	single stimulus continuous quality scale
SSD	sum of squared differences
STFM/LTFM	short-term frame memory/long-term frame memory

TDL	two-dimensional logarithmic
TMN	test model near-term
TR	temporal replacement
TSS	three-steps search
TV	television
UMTS	universal mobile telecommunication system
VD	vertical difference
VLC	variable-length coding
VO	video object
VOL	video object layer
VOP	video object plane; visual object plane
VQ	vector quantization
VS	video session
WBA	warping-based algorithm
WG	working group
WHT	Walsh-Hadamard transform

Chapter 1

Introduction to Mobile Video Communications

1.1 Motivations and Applications

In recent years, two distinct technologies have experienced massive growth and commercial success: *multimedia* and *mobile communications*. With the increasing reliance on the availability of multimedia information and the increasing mobility of individuals, there is a great need for providing multimedia information on the move. Motivated by this vision of being able to communicate from anywhere at any time with any type of information, a natural convergence of mobile and multimedia is under way. This new area is called *mobile multimedia communications*.

Mobile multimedia communications is expected to achieve unprecedented growth and worldwide success. For example, in Western Europe alone, it is estimated that by the year 2005 about 32 million people will use mobile multimedia services, representing a market segment worth 24 billion Euros per year and generating 3,800 million Mbytes of traffic per month. This will correspond, respectively, to 16% of all mobile users, 23% of the total revenues, and 60% of the overall traffic. Usage is expected to increase at even higher rates, with 35% of all mobile users having mobile multimedia services by the year 2010 [1]. The estimates become even more impressive when put in the context of a worldwide mobile market that reached 331.5 million users by the end of June 2000 [2] and is expected to grow to 1.7 billion users by 2010 [1]. It is not surprising, therefore, that this area has become an active and important research and development topic for both industry and academia, with groups across the world working to develop future mobile multimedia systems.

The definition of the term *multimedia* has always been a source of great debate and confusion. In this book, it refers to the presentation of information through multiple forms of media. This includes textual media (text, style, layout, etc.), numerical media (spreadsheets, databases, etc.), audio media (voice, music, etc.), visual media (images, graphics, video, etc.) and any other form of information representation.

Current second-generation mobile communication systems, like the *Global System for Mobile* (GSM),[1] already support a number of basic multimedia communication services. Examples are voice, basic fax/data, short message services, information-on-demand (e.g., sports results, news, weather), e-mail, still-image communication, and basic internet access. However, many technologically demanding problems need to be solved before real-time *mobile video communication* can be achieved. When such challenges are resolved, a wealth of advanced services and applications will be available to the mobile user. Examples are:

- Video-on-demand.

- Distance learning and training.

- Interactive gaming.

- Remote shopping.

- Online media services, such as news reports.

- Videotelephony.

- Videoconferencing.

- Telemedicine for remote consultation and diagnosis.

- Telesurveillance.

- Remote consultation or scene-of-crime work.

- Collaborative working and telepresence.

1.2 Main Challenges

The primary focus of this book is *mobile video communication*. In particular, the book focuses on three main challenges:

1. **Higher coding efficiency**. The radio spectrum is a very limited and scarce resource. This puts very stringent limits on the bandwidth available for a mobile channel. Given the enormous amount of data generated

---

[1]Originally, GSM was an acronym for *Group Spécial Mobile*.

by video, the use of efficient coding techniques is vital. For example, real-time transmission of a CIF[2] video at 15 frames/s over a 9.6 kbits/s GSM channel requires a compression ratio of about 1900:1. Although current coding techniques are capable of providing such compression ratios, there is a need for even higher coding efficiency to improve the quality (i.e., larger formats, higher frame rates, and better visual quality) of video at such very low bit rates. This continues to be the case even with the introduction of enhancements to second-generation systems, like the *General Packet Radio Service* (GPRS) [3] and the *Enhanced Data Rates for GSM Evolution* (EDGE), and also with the deployment of future higher-capacity, third-generation systems, like the *Universal Mobile Telecommunication System* (UMTS) [4].

2. **Reduced computational complexity.** In mobile terminals, processing power and battery life are very limited and scarce resources. Given the significant amount of computational power required to process video, the use of reduced-complexity techniques is essential. For example, recent implementations of video codecs [5,6] indicate that even state-of-the-art *digital signal processors* (DSPs) cannot, yet, achieve real-time video encoding. Typical results quoted in Refs. 5 and 6 are 1–5 frames/s using small video formats like SQCIF and QCIF.[3]

3. **Improved error resilience.** The mobile channel is a hostile environment with high bit error rates caused by a number of loss mechanisms, like multipath fading, shadowing, and co-channel interference. In the case of video, the effects of such errors are magnified due to the fact that the video bitstream is highly compressed to meet the stringent bandwidth limitations. In fact, the higher the compression is, the more sensitive the bitstream is to errors, since in this case each bit represents a larger amount of decoded video. The effects of errors on video are also magnified by the use of predictive coding and variable-length coding (VLC). The use of such coding methods can lead to temporal and spatial error propagation. It is, therefore, not difficult to realize that when transmitted over a mobile channel, compressed video can suffer severe degradation and the use of error-resilience techniques is vital.

---

[2]CIF stands for *Common Intermediate Format.* It is a digital video format in which the luminance component is represented by 352 pels × 288 lines and the two chrominance components are each of dimensions 176 × 144, where each pel is usually represented by 8 bits. Digital video formats are discussed in more detail in Chapter 2.

[3]*Quarter-CIF* (QCIF) has a luminance component of 176 × 144, whereas *sub-QCIF* (SQCIF) has a luminance component of 128 × 96.

It should be emphasized that those are not the only requirements of a mobile video communication system. Requirements like low delay, interactivity, scalability, and security are equally important.

1.3 Possible Solutions

Mobile video communication is a truly interdisciplinary subject [7]. Complete systems are likely to draw together solutions from different areas, like video source coding, channel coding, network design, semiconductor design, and others. This book will concentrate on solutions based on the *video source coding* part of the area. Thus, before being able to present the adopted approach, a closer look at video source coding is in order. Figure 1.1 shows a typical video codec. Changes between consecutive frames of a video sequence are mainly due to the movement of objects. Thus, the *motion estimation* (ME) block uses a motion model to estimate the movement that occurred between the current frame and a reference frame (usually a previously decoded frame that was stored in a frame buffer). This motion information is then utilized by the *motion compensation* (MC) block to move the contents of the reference frame to provide a prediction of the current frame. This *motion-compensated prediction* (MCP) is also known as the *displaced frame* (DF). The prediction is then subtracted from the current frame to produce an error signal known as

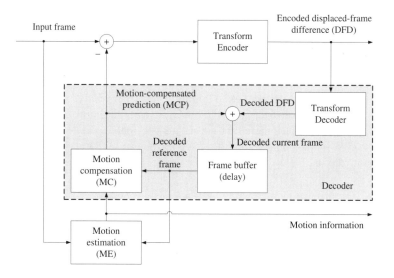

Figure 1.1: Typical video codec

the *displaced-frame difference* (DFD). Instead of encoding the current frame itself, this error signal is encoded, since it has a much reduced entropy. At the decoder, the same reference frame is used along with the received motion information to produce the same prediction. This prediction is then added to the received error signal to reconstruct the current frame.

Careful examination of this codec (as will be detailed in subsequent chapters) reveals that a *motion-based approach* can be adopted to provide suitable solutions for the three challenges of higher coding efficiency, reduced complexity, and error resilience. This motion-based approach can be summarized as follows:

1. **Advanced motion estimation techniques**. One way to achieve higher coding efficiency is to improve the performance of the motion estimation and compensation processes. The aim is to produce a better motion-compensated prediction and consequently reduce the entropy of the DFD signal. This should be achieved at the same or, preferably, a reduced motion overhead.[4]

2. **Reduced-complexity motion estimation techniques**. Motion estimation is the most computationally intensive process in a typical video codec. In fact, profiling results (as will be shown in Chapter 7) indicate that the computational complexity of this process is greater than that of all the remaining encoding steps combined. Thus, by reducing the complexity of this process, the overall complexity of the codec can be reduced.

3. **Motion-compensated error concealment techniques**. Apart from control and header data, the output of a typical video codec is one of two types: motion data or error (i.e., DFD) data.[5] Among the two types, motion data carries, in general, most of the information about a frame. In fact, at very low bit rates (typical of mobile video communication), motion data consumes a very high percentage of the available bit budget [8]. Thus, in the case of errors, it is very important to recover lost or erroneously received motion information. A class of error-resilience techniques that achieves this is motion-compensated error concealment, also known as *temporal error concealment*. Such techniques are particularly suited for mobile video communication, since, unlike other error resilience techniques, they do not increase the bit rate and they do not introduce any delay.

---

[4]An increase in motion overhead can be tolerated provided that the overall rate-distortion performance is improved.

[5]In the case of intracoded frames, the error signal is the same as the frame signal and no motion data is transmitted.

Part I

Introduction to Video Coding

This part gives an introduction to video coding. It contains two chapters. Chapter 2 introduces some of the fundamentals of video source coding. It starts by giving some basic definitions and then covers both analog and digital video along with some basic video coding techniques. It also presents the performance measures and the test sequences that will be used throughout the book. It then reviews both intraframe and interframe video coding methods.

Chapter 3 gives a brief introduction to video coding standards. The chapter starts by highlighting the need for video coding standards. It then outlines the chronological development of video coding standards, highlighting their main techniques and targeted applications. The chapter then concentrates on H.263 (and its recent extensions: H.263+ and H.263++) and MPEG-4 as examples of the state-of-the-art video coding standards.

Chapter 2

Video Coding: Fundamentals

2.1 Overview

This chapter gives a brief introduction to some fundamentals of video coding. Many of the concepts introduced in this chapter will be referenced and used in subsequent chapters. Section 2.2 gives some definitions. Section 2.3 covers analog video, whereas Section 2.4 concentrates on digital video. Section 2.5 introduces some of the basics of video coding. It also presents the performance measures and the test sequences that will be used in this book. Section 2.6 reviews intraframe video coding methods, whereas Section 2.7 reviews interframe coding methods.

2.2 What Is Video?

A *still image* is a spatial distribution of intensity[1] that is constant with respect to time [10]. *Video*, on the other hand, is a spatial intensity pattern that changes with time. Another common term for video is *image sequence*, since video can be represented by a time sequence of still-images.

2.3 Analog Video

2.3.1 Analog Video Signal

Video has traditionally been captured, stored, and transmitted in analog form. The term *analog video signal* refers to a one-dimensional (1-D) electrical

---

[1]Intensity is a measure over some interval of the electromagnetic spectrum of the flow of power that is radiated from, or incident on, a surface. It is usually measured in watts per square meter [9].

signal of time that is obtained by sampling the video intensity pattern in the *vertical* and *temporal* coordinates and converting intensity to electrical representation. This sampling process is known as *scanning*.

Raster scanning begins at the top-left corner and progresses horizontally, with a slight slope vertically, across the image. When it reaches the right-hand edge it snaps back to the left-hand edge (*horizontal retrace*) to start a new scan line. On reaching the bottom-right corner, a complete *frame* has been scanned and scanning snaps back to the top-left corner (*vertical retrace*) to begin a new frame. During retrace, blanking (black) and synchronization pulses are inserted.

The most commonly used raster scanning methods are progressive and interlaced, as illustrated in Figure 2.1. In *progressive* (also known as *noninterlaced* or 1:1) *scanning*, a frame is formed by a single scanning pass. In *interlaced* (or 2:1) *scanning*, however, a frame is formed by two successive scanning passes. In the first pass, the odd lines are scanned to form the first *field*, then the even lines are scanned to form the second field. When interleaved, the lines of the two fields form a single frame.

The aspect ratio, vertical resolution, frame rate, and refresh rate are important parameters of the video signal. The *aspect ratio* is the ratio of the width to the height of a frame. The *vertical resolution* is related to the number of scan lines per frame (including the blanking intervals). The *frame rate* is the number of frames scanned per second. The effect of smooth motion can be achieved using a frame rate of about 25–30 frames/s. However, at these frame rates the human eye picks up the flicker produced by refreshing the display between frames. To avoid this, the display refresh rate must be above 50 Hz.

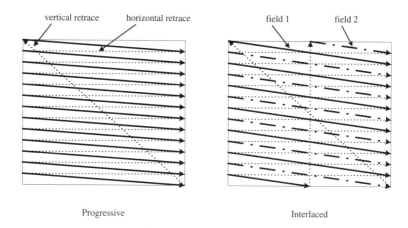

Figure 2.1: Raster scanning methods

Different industries employ different combinations of video parameters. For example, the computer industry uses progressive scanning with a frame rate of 72 frames/s. To reduce bandwidth requirements, the television industry uses interlaced scanning. In this case, the field rate is set to 50 or 60 fields/s to avoid refresh flicker,[2] while the frame rate (which, in interlaced video, is half the field rate) is 25 or 30 frames/s to maintain smooth motion. Note that this saving in bandwidth is at the expense of vertical resolution. There are two main television scanning systems: 625/50 (625 scan lines and 50 fields/s) and 525/60.

2.3.2   Color Representation

The preceding discussion considered monochrome video. In practice, however, most videos are in color. According to the *trichromatic theory* of color vision [11], color is perceived via *three* classes of cone cells, or photoreceptors, in the eye. Consequently, a color video can be produced by the superposition of three video signals. Each signal represents one of the three *primary* colors: red, green, and blue (RGB).[3] Practical television (TV) and video systems usually convert this RGB representation to a different color space of *luminance*[4] (which is closely related to the perception of brightness[5]) and *chrominance* (which is related to the perception of color hue[6] and saturation[7]). This representation serves two purposes. First, luminance ensures backward compatibility with monochrome video. Second, this representation lends itself more easily to video compression. This can be explained as follows. The human visual system (HVS) has poor response to color (chrominance) spatial detail compared to its response to luminance spatial detail [9]. Thus, the chrominance signals can be bandlimited or subsampled to achieve compression.

There are three main analog color coding systems: *Phase Alternation Line* (PAL), *SEquential Couleur Avec Memoire* (SECAM) and *National Television System Committee* (NTSC). They differ mainly in the way they

---

[2]Originally, television refresh rates were chosen to match the local AC power line frequency.

[3]The RGB is an *additive* color system. This means that when all the primaries are *added* in equal maximum quantities, the color white is perceived. In printing and painting, the cyan, magenta, and yellow (CMY) system is used. This is a *subtractive* color system since the total absorbtion of all three primaries produces the color white.

[4]Luminance is proportional to the light energy emitted per unit area of the source, but this energy is weighted according to the spectral sensitivity of the eye [9].

[5]Brightness is the attribute of a visual sensation according to which an area appears to emit more or less light [9].

[6]Hue is the attribute of a visual sensation according to which an area appears to be similar to one of the perceived colors, red, yellow, green and blue, or a combination of two of them [9].

[7]Saturation is the colorfulness of an area judged in proportion to its brightness [9].

calculate the luminance/chrominance components from the RGB components. For example, the PAL system calculates the luminance/chrominance components as follows:

$$Y' = +0.299R' + 0.587G' + 0.114B',$$

$$U' = -0.147R' - 0.289G' + 0.436B' = 0.493(B' - Y'), \qquad (2.1)$$

$$V' = +0.615R' - 0.515G' - 0.100B' = 0.877(R' - Y'),$$

where $R'G'B'$ are *gamma-corrected*[8] components in the range $[0,1]$. Note that Y' is closely related to gamma-corrected luminance and is usually referred to as *luma*. The chrominance is calculated as two *color-difference* components U' and V'. Again, since they are gamma-corrected, they are referred to as *chroma* components. The NTSC and SECAM systems calculate luma in the same way but use different coefficients for obtaining the chroma components (I' and Q' in NTSC, and D'_B and D'_R in SECAM).

2.3.3   Analog Video Systems

There are three main analog video systems. In most of Western Europe, a $625/50$ PAL system is used. In Russia, France, the Middle East, and Eastern Europe, a $625/50$ SECAM system is used. In North America and Japan, a $525/60$ NTSC system is used. All three systems are interlaced with a $4:3$ aspect ratio.

The three systems are *composite*. This means that the chroma components are first bandlimited and then combined (for example, by frequency interleaving) with the luma component. The resulting composite video signal has the same bandwidth as the original luma signal. For example, in the $625/50$ PAL system, the luma signal has a bandwidth of $5.5\,\text{MHz}$. The chroma signals are bandlimited to about $1.5\,\text{MHz}$ and then QAM (quadrature amplitude modulation) modulated with a color subcarrier at $4.43\,\text{MHz}$ above the picture carrier. For a more detailed discussion of these systems the reader is referred to Ref. 13. There are also other analog video systems that use separate components (component video) or a separate luma component and a composite chroma component (S-video) [10].

---

[8]In a video system, it is important to convey luminance in such a way that noise and quantization have a perceptually similar effect across the entire scale from black to white. This is achieved by applying a nonlinear function to each of the linear RGB components. This process is known as *gamma-correction* [12].

2.4 Digital Video

2.4.1 Why Digital?

For the past two decades or so, the world has been experiencing a digital revolution. Most industries have witnessed a change from analog to digital technology, and video was no exception. Digital video has the following advantages over analog video

- Ease of editing, enhancing, and creating special effects.

- Avoidance of artefacts typical of analog video, like, for example, those caused by repeated recording on tapes, and errors in color rendition due to inaccuracies in the separation of composite video signals.

- Easy software conversion from one standard to another. For analog video conversion, expensive transcoders are needed.

- Robustness to noise and ease of encryption.

- Ease of scalability (spatial, temporal, or signal-to-noise ratio (SNR)). This facilitates the provision of the same service over a wide range of networks and hardware platforms.

- Interactivity.

- Ease of indexing, search and retrieval. For analog video, this requires tedious visual scanning.

These advantages allowed a number of new applications and services to be introduced. For example, the TV broadcasting industry is introducing new services like interactivity, search and retrieval, video-on-demand, and high-definition television (HDTV). The telecommunication industry is providing videoconferencing and videophones over a wide range of wired and wireless networks. The computer industry is providing desktop video and videoconferencing. Other applications include intelligent highway traffic control systems, medical imaging, surveillance, and flight simulation, to mention a few.

2.4.2 Digitization

The process of digitizing video involves three basic operations: *filtering*, *sampling*, and *quantization*. If the frequency content of the input analog signal exceeds half the sampling frequency, aliasing artefacts will occur. Thus, the filtering operation is used to bandlimit the input signal and condition it for the following sampling operation.

The amplitude of the filtered analog signal is then sampled at specific time instants to generate a discrete-time signal. The minimum sampling rate is known as the *Nyquist rate* and is equal to twice the signal bandwidth.

The resulting discrete-time samples have continuous amplitudes. Thus, it would require infinite precision to represent them. The quantization operation is used to map such values onto a finite set of discrete amplitudes that can be represented by a finite number of bits.

Each discrete-time, discrete-amplitude sample is called a *picture element* and is usually abbreviated to a *pel* or a *pixel*. The pels are arranged in a two-dimensional (2-D) array to form a digital still image or a digital frame. A digital video consists of a sequence of such digital frames.

For color video, the foregoing operations are repeated for each component. Thus, a digital still image would normally be represented by three 2-D arrays. Almost all digital video systems use component representation. This avoids the artefacts that result from composite encoding.[9]

As an example, consider the digitization of a 625/50 PAL analog signal. The luma and chroma components are first filtered to 5.5 MHz and 1.5 MHz, respectively. During sampling, minimum sampling frequencies of 11 MHz and 3 MHz must be used to sample the luma and chroma components, respectively. The resulting discrete-time signals are then quantized to a given precision (usually 8 bits).

2.4.3   Chroma Subsampling

As already mentioned, the HVS has poor response to chrominance spatial detail compared to its response to luminance spatial detail. This property can be exploited to reduce bandwidth requirements by subsampling the chroma components. The most commonly used subsampling patterns are illustrated in Figure 2.2. In 4:2:2 subsampling, the chroma components are subsampled by a factor of 2 horizontally. This gives a reduction of about 33% in the overall raw data rate. In 4:1:1 subsampling, the chroma components are subsampled by a factor of 4 horizontally, giving a reduction of 50%. In 4:2:0 subsampling, the chroma components are subsampled by a factor of 2 both horizontally and vertically, giving a reduction of 50% in the overall raw data rate. Vertically subsampled chroma samples are always sited midway between luma samples. Horizontally subsampled chroma samples, however,

---

[9]As already discussed, composite encoding is used in analog systems to save bandwidth. In digital systems, however, bandwidth is saved using digital video compression techniques, as will be described later.

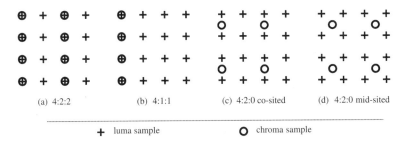

Figure 2.2: Chroma subsampling patterns

can be either midway between luma samples (Figure 2.2(d)) or co-sited with odd-numbered luma samples (Figure 2.2(c)).[10]

2.4.4 Digital Video Formats

Exchange of digital video between different industries, applications, networks, and hardware platforms requires standard digital video formats. Following are the most commonly used formats.

2.4.4.1 CCIR-601

The *International Consultative Committee for Radio* (CCIR)[11] Recommendation 601 [14] defines a digital video format for the international exchange and broadcast of production-quality TV programs. As with analog standards, CCIR-601 defines two interlaced systems: 525/60 and 625/50. The main family within the standard uses a chroma subsampling of 4:2:2. The luma sampling frequency is 13.5 MHz, the chroma sampling frequency is $13.5 \times 0.5 = 6.75$ MHz, and the components are quantized to 8 bits. In the 525/60 system, the luma component of the frame has active dimensions of 720 pels $\times$ 480 lines and the chroma components have 360 pels $\times$ 480 lines. In the 625/50 system, the corresponding values are 720×576 for luma and 360×576 for chroma. Note that despite the differences between the two systems, they generate the same raw bit rate[12] of 165.89 Mbits/s. The standard is based on component

---

[10]In view of this lack of consistency, the authors adopt the terms *mid-sited* and *co-sited* to describe the two cases.

[11]The CCIR is currently known as ITU-R (*International Telecommunications Union—Radio Sector*).

[12]Bit rate $= [(720 \times 480) + 2(360 \times 480)] \times 30 \times 8 = [(720 \times 576) + 2(360 \times 576)] \times 25 \times 8 = 165888000$ bits/s, where the 2 refers to the two chroma components, the 30 and the 25 are the frame rates of the two systems, and the 8 is the number of bits per sample.

video with one luma (Y') and two chroma $(C_R'$ and $C_B')$ components calculated as follows:

$$Y' = 219(+0.299R' + 0.587G' + 0.114B') + 16,$$

$$C_B' = 224(-0.169R' - 0.331G' + 0.500B') + 128, \qquad (2.2)$$

$$C_R' = 224(+0.500R' - 0.419G' - 0.081B') + 128,$$

where Y' has 220 levels in the range [16, 235], with black at 16 and white at 235, and C_B' and C_R' have 225 levels in the range [16, 240], with zero difference at 128. Note that other levels within the 8-bit range [0, 255] are reserved for synchronization and signal processing head- and foot-rooms.

2.4.4.2 SIF and QSIF

CCIR-601 was defined mainly for broadcast-quality applications. For storage applications, a lower-resolution format called the *Source Input Format* (SIF) was defined. This is a progressive 4:2:0 mid-sited format with a luma component that is half the CCIR-601 active luma component in both dimensions. The CCIR-601 format has 720 luminance pels/line, which means that an SIF format must have $720/2 = 360$ luma pels/line. Since 360 is not divisible by 16 (which is the main coding unit within standard video codecs), 8 pels (4 from each side) are usually discarded to reduce the number of luma pels per line to 352. Since there are two CCIR-601 systems, there are two SIF formats: the first has a luma component of 352×240, chroma components of 176×120, and a frame rate of 30 frames/s, whereas the second format has a luma of 352×288, chromas of 176×144, and a frame rate of 25 frames/s.

A lower-resolution version of SIF is the *quarter-SIF* (QSIF) format. It has half the dimensions of SIF in both directions. This means it has quarter the number of samples, hence the name. Again, two versions are available: the first has a luma of 176×120, chromas of 88×60, and a frame rate of 30 frames/s, whereas the second has a luma of 176×144, chromas of 88×72, and a frame rate of 25 frames/s. For methods of converting between CCIR-601, SIF and QSIF, refer to Ref. 15.

2.4.4.3 CIF and Its Family

In order for video codecs to cope with both 525/60 and 625/50 formats, a common format was defined. In this format, the luma component has a horizontal resolution that is half that of both CCIR-601 systems, a vertical resolution that is half that of the 625/50 system, and a temporal resolution that is half that of the 525/60 system. This intermediate choice of vertical resolution from one system and temporal resolution from the other leads to the name *Common Intermediate Format* (CIF). The CIF is progressive, with

Table 2.1: The CIF family

	Luma		Chromas	
	pels/line	lines/frame	pels/line	lines/frame
SQCIF	128	96	64	48
QCIF	176	144	88	72
CIF	352	288	176	144
4CIF	704	576	352	288
16CIF	1408	1152	704	576

4:2:0 mid-sited chroma subsampling and a frame rate of 30 frames/s. There are a number of lower- and higher-resolution members in the CIF family. Those are defined in Table 2.1.

2.4.4.4 Other Formats

There are a number of other formats. For example, some HDTV systems[13] use a 1440×1050 luma at 30 frames/s with progressive scanning and no chroma subsampling (i.e., 4:4:4).

2.5 Video Coding Basics

2.5.1 The Need for Video Coding

Table 2.2 shows the raw data rates of a number of typical video formats, whereas Table 2.3 shows a number of typical video applications and the bandwidths available to them. It is immediately evident that video coding (or *compression*) is a key enabling technology for such applications. Consider a 2-hour CCIR-601 color movie. Without compression, a 5-Gbit compact disc (CD) can hold only 30 seconds of this movie. To store the entire movie on the same CD requires a compression ratio of about 240:1. Without compression, the same movie will take about 36 days to arrive at the other end of a 384 kbits/s Integrated Services Digital Network (ISDN) channel. To achieve real-time transmission of the movie over the same channel, a compression ratio of about 432:1 is required.

---

[13] A range of HDTV formats exist.

Table 2.2: Raw data rates of typical video formats

Format	Raw data rate
HDTV	1.09 Gbits/s
CCIR-601	165.89 Mbits/s
CIF @ 15 f.p.s.	18.24 Mbits/s
QCIF @ 10 f.p.s.	3.04 Mbits/s

Table 2.3: Typical video applications

Application	Bandwidth
HDTV (6-MHz channel)	20 Mbits/s
Desktop video (CD-ROM)	1.5 Mbits/s
Videoconferencing (ISDN)	384 kbits/s
Videophone (PSTN)	56 kbits/s
Videophone (GSM)	10 kbits/s

2.5.2   Elements of a Video Coding System

The aim of video coding is to reduce, or compress, the number of bits used to represent video. Video signals contain three types of redundancy: statistical, psychovisual, and coding redundancy. *Statistical redundancy* is present because certain data patterns are more likely than others. This is mainly due to the high spatial (*intraframe*) and temporal (*interframe*) correlations between neighboring pels. *Psychovisual redundancy* is due to the fact that the HVS is less sensitive to certain visual information than to other visual information. If video is coded in a way that uses more and/or longer code symbols than absolutely necessary, it is said to contain *coding redundancy*. Video compression is achieved by reducing or eliminating these redundancies.

Figure 2.3 shows the main elements of a video encoder. Each element is designed to reduce one of the three basic redundancies.

The *mapper* (or *transformer*) transforms the input raw data into a representation that is designed to reduce statistical redundancy and make the data more amenable to compression in later stages. The transformation is a one-to-one mapping and is, therefore, reversible.

Figure 2.3: Elements of a video encoder

The *quantizer* reduces the accuracy of the mapper's output, according to some fidelity criterion, in an attempt to reduce psychovisual redundancy. This is a many-to-one mapping and is, therefore, irreversible.

The *symbol encoder* (or *codeword assigner*) assigns a codeword, a string of binary bits, to each symbol at the output of the quantizer. The code must be designed to reduce coding redundancy. This operation is reversible.

In general, compression methods can be classified into lossless methods and lossy methods. In *lossless* methods the reconstructed (compressed-decompressed) data is identical to the original data. This means that such methods do not employ a quantizer. Lossless methods are also known as *bit-preserving* or *reversible* methods. In lossy methods the reconstructed data is not identical to the original data; that is, there is loss of information due to the quantization process. Such methods are therefore *irreversible*, and they usually achieve higher compression than lossless methods.

2.5.3   Elements of Information Theory

A source $\mathcal{S}$ with an alphabet $\mathcal{A}$ can be defined as a discrete random process $\mathcal{S} = S_1, S_2, \ldots$, where each random variable S_i takes a value from the alphabet $\mathcal{A}$.

In a *discrete memoryless source* (DMS) the successive symbols of the source are statistically independent. Such a source can be completely defined by its alphabet $\mathcal{A} = \{a_1, a_2, \ldots, a_N\}$ and the associated probabilities $\mathcal{P} = \{p(a_1), p(a_2), \ldots, p(a_N)\}$, where $\sum_{i=1}^{N} p(a_i) = 1$. According to information theory, the information I contained in a symbol a_i is given by

$$I(a_i) = \log_2 \frac{1}{p(a_i)} = -\log_2 p(a_i) \quad \text{(bits)}, \qquad (2.3)$$

and the average information per source symbol $H(\mathcal{S})$, also known as the *entropy* of the source, is given by

$$H(\mathcal{S}) = \sum_{i=1}^{N} p(a_i) I(a_i) = -\sum_{i=1}^{N} p(a_i) \log_2 p(a_i) \quad \text{(bits/symbol)}. \qquad (2.4)$$

A more realistic approach is to model sources using Markov-K random processes. In this case the probability of occurrence of a symbol depends on the values of the K preceding symbols. Thus, a Markov-K source can be specified by the conditional probabilities $p(S_j = a_i | S_{j-1}, \ldots, S_{j-K})$, for all j, $a_i \in \mathcal{A}$. In this case, the entropy is given by

$$H(\mathcal{S}) = \sum_{S^K} p(S_{j-1}, \ldots, S_{j-K}) H(\mathcal{S} | S_{j-1}, \ldots, S_{j-K}), \qquad (2.5)$$

where S^K denotes all possible realizations of $S_{j-1}, \ldots, S_{j-K}$, and

$$H(\mathscr{S}|S_{j-1}, \ldots, S_{j-K})$$
$$= -\sum_{a_i \in \mathscr{A}} p(S_j = a_i|S_{j-1}, \ldots, S_{j-K}) \log p(S_j = a_i|S_{j-1}, \ldots, S_{j-K}).$$

(2.6)

The performance bound of a lossless coding system is given by the lossless coding theorem [16]:

> *Lossless coding theorem*: The minimum bit rate R_{min} that can be achieved by lossless coding of a source $\mathscr{S}$ can be arbitrarily close, but not less than, the source entropy $H(\mathscr{S})$. Thus $R_{min} = H(\mathscr{S}) + \varepsilon$, where ε is a positive quantity that can be made arbitrarily close to zero.

For a DMS, this lower bound can be approached by coding symbols independently, whereas for a Markov-K source, blocks of K symbols should be encoded at a time.

The performance bounds of lossy coding systems are addressed by a branch of information theory known as *rate-distortion theory* [16, 17, 18]. This theory provides lower bounds on the obtainable average distortion for a given average bit rate, or vice versa. It also promises that codes exist that approach the theoretical bounds when the code dimension and delay become large. An important theorem in this branch is the source coding theorem [17]:

> *Source coding theorem*: There exists a mapping from source symbols to codewords such that for a given distortion D, $R(D)$ bits/symbol are sufficient to achieve an average distortion that is arbitrarily close to D.

The function $R(D)$ is known as the *rate-distortion function*. It is a convex, continuous, and strictly decreasing function of D, as illustrated in Figure 2.4. This function is normally computed using numerical methods [18], although for simple source and distortion models it can be computed analytically. Although rate-distortion theory does not give an explicit method for constructing practical optimum coding systems, it gives very important hints about the properties of such systems.

2.5.4 Quantization

As already discussed, quantization is a key element of a video coding system. Quantization can be viewed as a many-to-one mapping. It represents a set of continuous-valued samples with a finite number of symbols. If each input sample is quantized independently, then the process is referred to as *scalar*

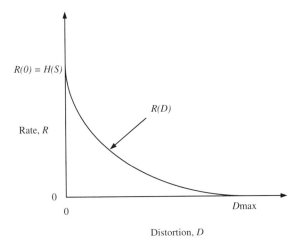

Figure 2.4: Rate-distortion function

quantization. If, however, the input samples are grouped into a set of vectors and this set is mapped to a finite number of vectors, then the process is known as *vector quantization*. Vector quantization is discussed in more detail in Section 2.6.4.

Assume that the quantizer input s varies between $s_{\min}$ and $s_{\max}$ and that this range is to be mapped to a finite set of N symbols, then a set of $N + 1$ *decision levels* d_i, $0 \leq i \leq N$, are first defined, where $d_0 = s_{\min}$ and $d_N = s_{\max}$. This divides the input range into N *quantization intervals*. At the output of the quantizer, each quantization interval is then represented by a *reconstruction level* r_i, $1 \leq i \leq N$. Thus, a scalar quantizer $Q(\cdot)$ can be defined as follows:

$$\dot{s} = Q(s) = r_i, \qquad \text{if } d_{i-1} < s \leq d_i, \quad \text{where } 1 \leq i \leq N, \qquad (2.7)$$

where $\dot{s}$ is the quantized output. There are, in general, two types of optimum scalar quantizers: Lloyd-Max and entropy-constrained. Lloyd-Max [19, 20] quantizers are designed to minimize the mean squared error with a fixed number of levels. Entropy-constrained quantizers [21] are designed to minimize a distortion measure for a constant output entropy.

The simplest form of scalar quantization is *uniform quantization*. In this case, the decision levels (and the reconstruction levels) are equally spaced, with a *quantizer step size* θ. In addition, the reconstruction levels are set to the midpoints of the quantization intervals. Figure 2.5(a) shows an example of a uniform quantizer, with $N = 7$ reconstruction levels. In this case,

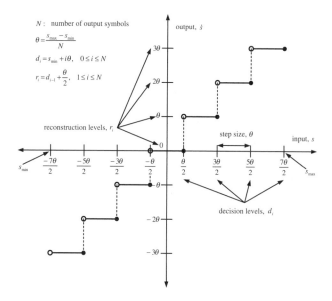

(a) Without dead zone

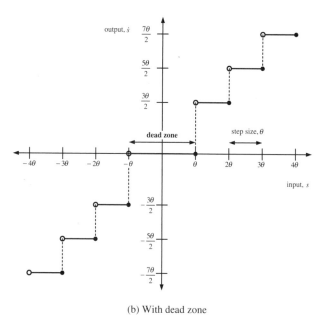

(b) With dead zone

Figure 2.5: Uniform threshold quantizers

the quantization process can be implemented at the encoder using

$$\hat{s} = \text{NINT}\left[\frac{s}{\theta}\right], \tag{2.8}$$

where NINT[·] is the operation of rounding to the nearest integer and $\hat{s}$ is called the *quantization index*. It is the quantization index that is encoded and sent to the decoder. The decoder can then dequantize this index to obtain the reconstructed output as follows:

$$\dot{s} = \theta \cdot \hat{s}. \tag{2.9}$$

This type of quantizer is also known as a *threshold quantizer*, because it quantizes to zero all those inputs whose magnitudes are below a threshold. As will be discussed later, this type of quantizer is usually used in transform coding to reduce the number of transform coefficients that need to be encoded. Another example of uniform threshold quantizers is illustrated in Figure 2.5(b). In this case, the quantization interval around zero has been extended to form a *dead zone*. This causes more nonsignificant inputs to be quantized to zero and, thus, increases compression. The quantization equation for this quantizer is given by

$$\hat{s} = \text{FIX}\left[\frac{s}{\theta}\right], \tag{2.10}$$

where FIX[·] is the operation of rounding to the nearest integer toward zero (i.e., truncation). The corresponding dequantization equation is given by

$$\dot{s} = \theta \cdot \hat{s} + \text{SIGN}(\hat{s}) \cdot \frac{\theta}{2}, \tag{2.11}$$

$$\text{SIGN}(a) = \begin{cases} +1, & a > 0, \\ 0, & a = 0, \\ -1, & a < 0. \end{cases} \tag{2.12}$$

Scalar quantizers can also be *nonuniform*. In this case, more reconstruction levels are assigned to more significant subintervals within the input range. This yields a higher overall accuracy.

2.5.5 Symbol Encoding

Another key element of video coding systems is the symbol encoder. This assigns a codeword to each symbol at the output of the quantizer. The symbol encoder must be designed to reduce the coding redundancy present in the set of symbols. Following are a number of commonly used techniques that can be applied individually or in combinations.

2.5.5.1 Run-Length Encoding

The output of the quantization step may contain long runs of identical symbols. One way to reduce this redundancy is to employ *run-length encoding* (RLE). There are different forms of RLE. For example, if the quantizer output contains long runs of zeros, then RLE can represent such runs with intermediate symbols of the form (RUN, LEVEL). For example, a run of the form $0, 0, 0, 0, 0, 9$ can be represented by the intermediate symbol $(5, 9)$.

2.5.5.2 Entropy Encoding

The quantizer can be considered a DMS $\mathcal{Q}$ that can be completely specified by its alphabet $\mathcal{R} = \{r_1, r_2, \ldots, r_N\}$, where r_i are the reconstruction levels and the associated probabilities of occurrence $\mathcal{P} = \{p(r_1), p(r_2), \ldots, p(r_N)\}$. The information contained in a symbol $I(r_i)$ is given by Equation (2.3), whereas the entropy of the source $H(\mathcal{Q})$ is given by Equation (2.4).

Now consider a symbol encoder that assigns a codeword c_i of length $l(c_i)$ bits to symbol r_i. Then the average word length $\bar{L}$ of the code is given by

$$\bar{L} = \sum_{i=1}^{N} p(r_i) l(c_i) \qquad \text{(bits)}, \qquad (2.13)$$

and the efficiency (η) of the code is

$$\eta = \frac{H(\mathcal{Q})}{\bar{L}}. \qquad (2.14)$$

Thus, an optimal ($\eta = 1$) code must have an average word length that is equal to the entropy of the source; i.e., $\bar{L} = H(\mathcal{Q})$. Clearly, this can be achieved if each codeword length is equal to the information content of the associated symbol, that is, $l(c_i) = I(r_i)$. Since $I(r_i)$ is inversely proportional to $p(r_i)$ (from Equation (2.3)), then an efficient code must assign shorter codewords to more probable symbols, and vice versa. This is known as *entropy encoding* or *variable-length coding* (VLC) (as opposed to *fixed-length coding* (FLC)).

The most commonly used VLC is *Huffman coding* [22]. Given a finite set of symbols and their probabilities, Huffman coding yields the optimal[14] integer-length prefix[15] code. The basic principles of Huffman coding can be illustrated using the example given in Figure 2.6. In each stage, the two least probable symbols are combined to form a new symbol with a probability equal

---

[14]Huffman is optimal in the sense that no other integer-length VLC can achieve a smaller average word length.

[15]In a prefix code, no codeword is a prefix of another codeword. This makes the code uniquely decodable.

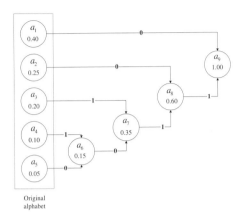

Original
alphabet

Figure 2.6: Huffman coding example

Table 2.4: Comparison between VLC (of Figure 2.6) and a 3-bit FLC

r_i	$p(r_i)$	$I(r_i)$	VLC c_i	FLC c_i
a_1	0.40	1.32 bits	0 (1 bit)	000
a_2	0.25	2.00 bits	10 (2 bits)	001
a_3	0.20	2.32 bits	111 (3 bits)	010
a_4	0.10	3.32 bits	1101 (4 bits)	011
a_5	0.05	4.32 bits	1100 (4 bits)	100

$H(R) \approx 2.04$ bits/symbol
$\bar{L}_{FLC} = 3$ bits/word
$\eta_{FLC} \approx 0.68$

$\bar{L}_{VLC} \approx 2.1$ bits/word
$\eta_{VLC} \approx 0.97$

to the sum of their probabilities. This new symbol creates a new node in the tree, with two branches connecting it to the original two nodes. A "0" is assigned to one branch and a "1" is assigned to the other. The original two nodes are then removed from the next stage. This process is continued until the new symbol has a probability of 1. Now, to find the codeword for a given symbol, start at the right-hand end of the tree and follow the branches that lead to the symbol of interest combining the "0"s and "1"s assigned to the branches. Table 2.4 shows the obtained VLC and compares it to an FLC of 3 bits. Clearly, the Huffman VLC is much more efficient than the FLC.

There are more efficient implementations of Huffman coding. For example, in many cases, most of the symbols of a large symbol set have very small probabilities. This leads to very long codewords and consequently to large storage requirements and high decoding complexity. In the *modified Huffman code* [23] the less probable symbols (and their probabilities) are lumped into a single symbol like ESCAPE. A symbol in this new ESCAPE category is coded using the VLC codeword for ESCAPE followed by extra bits to identify the actual symbol. Standard video codecs also use 2-D and 3-D versions of the Huffman code. For example, the H.263 standard (see Section 3.4) uses a 3-D Huffman code where three different symbols (LAST, RUN, LEVEL) are lumped into a single symbol (EVENT) and then encoded using one VLC codeword.

One disadvantage of the Huffman code is that it can only assign integer-length codewords. This usually leads to a suboptimal performance. For example, in Table 2.4, the symbol a_3 was represented with a 3-bit codeword, whereas its information content is only 2.32 bits. In fact, Huffman code can be optimal only if all the probabilities are integer powers of $1/2$. An entropy code that can overcome this limitation and approach the entropy of the source is *arithmetic coding* [24]. In Huffman coding there is a one-to-one correspondence between the symbols and the codewords. In arithmetic coding, however, a single variable-length codeword is assigned to a variable-length block of symbols.

2.5.6 Performance Measures

When evaluating the performance of a video coding system, a number of aspects need to be assessed and measured. One important aspect is the amount of *compression* (C) achieved by the system. This can be measured in a number of ways:

$$C = \frac{\text{number of bits in original video}}{\text{number of bits in compressed video}} \quad \text{(unitless)}, \quad (2.15)$$

$$C = \frac{\text{number of bits in compressed video}}{\text{number of pels in original video}} \quad \text{(bits/pel)}, \quad (2.16)$$

$$C = \frac{\text{number of bits in compressed video}}{\text{number of frames in original video}} \times \text{frame rate} \quad \text{(bits/s)}. \quad (2.17)$$

Another important aspect is the *reconstruction quality*. This can be assessed using a number of subjective and objective measures. *Subjective measures* are normally evaluated by showing the reconstructed video to a group of subjects and asking for their views on the perceived quality. A number of subjective assessment methodologies have been developed over the years. Examples are the double stimulus impairment scale (DSIS) and the double and single stimulus continuous quality scales, (DSCQS) and (SSCQS), respectively. For a detailed description of such experiments the reader is referred to Ref. 25.

Despite their reliability, subjective quality experiments are expensive and time consuming. *Objective measures* provide cheaper and faster alternatives. One commonly used objective measure is the *mean squared error* (MSE), which is defined as

$$\text{MSE} = \frac{1}{H \times V} \sum_{x=1}^{H} \sum_{y=1}^{V} [f(x, y) - \hat{f}(x, y)]^2, \tag{2.18}$$

where H and V are the horizontal and vertical dimensions of the frame, respectively, and $f(x, y)$ and $\hat{f}(x, y)$ are the pel values at location (x, y) of the original and reconstructed frames, respectively. Care should be taken to include color components and to take into account any chroma subsampling. For example, the MSE of a reconstructed 4:2:0 color frame can be calculated as

$$\begin{aligned}
\text{MSE}_{4:2:0} = \frac{1}{\frac{3}{2}H \times V} &\left(\sum_{x=1}^{H} \sum_{y=1}^{V} [Y'(x, y) - \hat{Y}'(x, y)]^2 \right. \\
&+ \sum_{x=1}^{H/2} \sum_{y=1}^{V/2} [C_R'(x, y) - \hat{C}_R'(x, y)]^2 \\
&+ \left. \sum_{x=1}^{H/2} \sum_{y=1}^{V/2} [C_B'(x, y) - \hat{C}_B'(x, y)]^2 \right) \\
= \frac{2}{3}(\text{MSE}_{Y'} &+ \tfrac{1}{4}\text{MSE}_{C_R'} + \tfrac{1}{4}\text{MSE}_{C_B'}).
\end{aligned} \tag{2.19}$$

A more common form of the MSE measure is the *peak signal-to-noise ratio* (PSNR), which is defined as

$$\text{PSNR} = 10 \log_{10} \left(\frac{f_{\text{max}}^2}{\text{MSE}} \right) \qquad (\text{dB}), \tag{2.20}$$

where f_{max} is the maximum possible pel value (for example, 255 for an 8-bit resolution component). Although this measure does not always correlate well with perceived video quality, its relative simplicity makes it a very popular choice in the video coding community. Thus, to facilitate comparisons with other algorithms reported in the literature, this book adopts the PSNR measure. If accuracy is a major concern, then more sophisticated objective measures based on perceptual models can be used [26].

When testing a video coding algorithm, it is very important to subject it to a range of input video sequences with different characteristics and a reasonable

(a) FOREMAN   (b) AKIYO   (c) TABLE TENNIS

Figure 2.7: Three test sequences

spread of data properties. The *Moving Picture Experts Group* (MPEG) established a library of CCIR-601 test sequences divided into five classes: class A
(low spatial detail and low amount of motion), class B (medium spatial detail
and low amount of motion or vice versa), class C (high spatial detail and
medium amount of motion, or vice versa), class D (stereoscopic), and class E
(hybrid of natural and synthetic content) [27]. The first three classes are more
relevant to the work carried out in this book. Thus, the book uses three test
sequences: AKIYO, FOREMAN, and TABLE TENNIS, where each sequence is a representative of one of the three relevant classes, A, B, and C, respectively. The
three sequences are at QSIF resolution and include 300 frames each. This resolution is typical of the sequences used in very-low-bit-rate applications. Both
AKIYO and TABLE TENNIS have luma components of 176×120 and a frame rate
of 30 frames/s, whereas FOREMAN has a luma component of 176×144 and a
frame rate of 25 frames/s. Figure 2.7 shows the luma component of the first
frame of each of the three test sequences.

2.6   Intraframe Coding

Intraframe coding refers to video coding techniques that achieve compression
by exploiting (reducing) the high spatial correlation between neighboring pels
within a video frame. Such techniques are also known as *spatial redundancy
reduction* techniques or *still-image coding* techniques.

2.6.1   Predictive Coding

Predictive coding was originally proposed by Cutler in 1952 [28]. In this
method, a number of previously coded pels are used to form a *prediction* of
the current pel. The *difference* between the pel and its prediction forms the
signal to be coded. Obviously, the better the prediction, the smaller the error

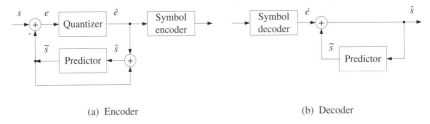

(a) Encoder   (b) Decoder

Figure 2.8: Block diagram of a predictive coding system

signal and the more efficient the coding system. At the decoder, the same prediction is produced using previously decoded pels, and the received error signal is added to reconstruct the current pel. A block diagram of a predictive coding system is depicted in Figure 2.8. Predictive coding is commonly referred to as *differential pulse code modulation* (DPCM). A special case of this method is *delta modulation* (DM), which quantizes the error signal using two quantization levels only.

Predictive coding can take many forms, depending on the design of the predictor and the quantizer blocks. The predictor can use a linear or a nonlinear function of the previously decoded pels, it can be 1-D (using pels from the same line) or 2-D (using pels from the same line and from previous lines), and it can be fixed or adaptive. The quantizer also can be uniform or nonuniform, and it can be fixed or adaptive.

The minimal storage and processing requirements were partly responsible for the early popularity of this method, when storage and processing devices were scarce and expensive resources. The method, however, provides only a modest amount of compression. In addition, its performance is highly dependent on the statistics of the input data, and it is very sensitive to errors (feedback through the prediction loop can cause error propagation). As processing and storage devices became more available, more complex, more efficient methods like transform coding have become more popular. Despite this, predictive coding is still used in video coding, as, for example, in the lossless coding of motion vectors.

2.6.2 Transform Coding

Transform coding, developed more than two decades ago, has proven to be a very effective video coding method. Today, it forms the basis of almost all video coding standards. Figure 2.9 shows a block diagram of a typical transform coding system. The input frame is first segmented into $N \times N$ blocks.

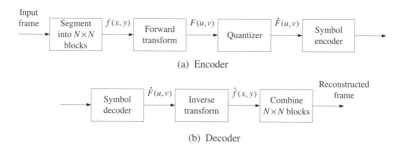

(a) Encoder

(b) Decoder

Figure 2.9: Block diagram of a transform coding system

A unitary[16] space-frequency transform is applied to each block to produce an $N \times N$ block of transform (spectral) coefficients that are then suitably quantized and coded. At the decoder, an inverse transform is applied to reconstruct the frame. The main goal of the transform is to decorrelate the pels of the input block. This is achieved by redistributing the energy of the pels and concentrating most of it in a small set of transform coefficients. This is known as *energy compaction*. The transform process can also be interpreted as a coordinate rotation of the input or as a decomposition of the input into orthogonal basis functions weighted by the transform coefficients [29]. Compression comes about from two main mechanisms. First, low-energy coefficients can be discarded with minimum impact on the reconstruction quality. Second, the HVS has differing sensitivity to different frequencies. Thus, the retained coefficients can be quantized according to their visual importance.

When choosing a transform, three main properties are desired: good energy compaction, data-independent basis functions, and fast implementation. The Karhunen-Loève transform (KLT) is the optimal transform in an energy-compaction sense. Unfortunately, this optimality is due to the fact that the KLT basis functions are dependent on the covariance matrix of the input block. Recomputing and transmitting the basis functions for each block is a nontrivial computational task. These disadvantages severely limit the use of the KLT in practical coding systems. The performance of many suboptimal transforms with data-independent basis functions have been studied [30]. Examples are the discrete Fourier transform (DFT), the discrete cosine transform (DCT), the Walsh-Hadamard transform (WHT), and the Haar transform. It has been demonstrated that the DCT has the closest energy-compaction performance to that of the optimum KLT [30]. This has motivated the development of a number of fast DCT algorithms, e.g., Ref. 31. Due to these

---

[16]A unitary transform is a reversible linear transform with orthonormal basis functions [29].

attractive features, i.e., near-optimum energy-compaction, data-independent basis functions and fast algorithms, the DCT has become the "workhorse" of most image and video coding standards.

The DCT was developed by Ahmed *et al.* in 1974 [32]. There are four slightly different versions of the DCT [33], but the one commonly used for video coding is denoted by DCT-II. The 2-D DCT-II of an $N \times N$ block of pels is given by

$$F(u,v) = C(u)C(v) \sum_{x=0}^{N-1} \sum_{y=0}^{N-1} f(x,y) \cos\left(\frac{(2x+1)u\pi}{2N}\right) \cos\left(\frac{(2y+1)v\pi}{2N}\right),$$

$$(2.21)$$

where $f(x,y)$ is the pel value at location (x,y) within the block, $F(u,v)$ is the corresponding transform coefficient, $0 \leq u,v,x,y \leq N-1$, and

$$C(\alpha) = \begin{cases} \sqrt{\frac{1}{N}}, & \alpha = 0, \\ \sqrt{\frac{2}{N}}, & \text{otherwise.} \end{cases}$$

$$(2.22)$$

The transform coefficient $F(0,0)$ at the top-left corner of the transformed block is called the DC coefficient because it contains the lowest frequencies in both the horizontal and vertical dimensions. The corresponding inverse DCT transform is given by

$$f(x,y) = \sum_{u=0}^{N-1} \sum_{v=0}^{N-1} C(u)C(v)F(u,v) \cos\left(\frac{(2x+1)u\pi}{2N}\right) \cos\left(\frac{(2y+1)v\pi}{2N}\right).$$

$$(2.23)$$

It can be deduced from Equation (2.21) that the computational complexity of an $N \times N$ 2-D DCT is of the order $\mathcal{O}(N^4)$. However, one of the advantages of the DCT is that it is separable. This means that a 2-D DCT can be separated into a pair of 1-D DCTs. Thus, to obtain the 2-D DCT of an $N \times N$ block, a 1-D DCT is performed first on each of the N rows of the block and then on each of the N columns of the resulting block (or vice versa). The same applies to the inverse DCT. This reduces the complexity to $\mathcal{O}(2N^3)$. Further reductions in complexity can be achieved using a number of fast DCT algorithms [31].

Beside transform selection, a significant factor that affects transform coding performance and computational complexity is the block size. In general, the

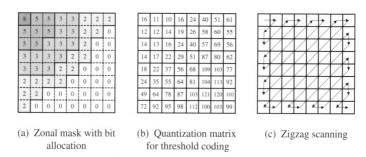

| (a) Zonal mask with bit allocation | (b) Quantization matrix for threshold coding | (c) Zigzag scanning |

Figure 2.10: Transform coefficient bit allocation

use of smaller block sizes reduces computational complexity.[17] However, as will be discussed later, transform coding suffers from blocking artefacts at very low bit rates. Such artefacts are more disturbing with smaller block sizes [15]. As a compromise between computational complexity and blocking artefacts, most transform coding systems employ a block size of 8×8 or 16×16. Note that both sizes are powers of 2, which simplifies computations.

Another important factor in transform coding is *bit allocation*. This refers to the process of determining which coefficients should be retained for coding and how coarsely each retained coefficient should be quantized. There are two main approaches: zonal coding and threshold coding. In *zonal coding* the retained coefficients are selected on the basis of maximum variance. Thus, the locations of the retained coefficients with the largest variances are indicated by a zonal mask that is the same for all blocks. Once the retained coefficients are decided, a number of methods can be used to decide the number of bits allocated to each. One method is to choose the number of bits to be proportional to the variance of the coefficient. Figure 2.10(a) shows a zonal mask with the allocated bits. Once the number of bits allocated for each coefficient is determined, a different quantizer can be designed for each coefficient.

One disadvantage of zonal coding is that the locations of the retained coefficients and the bits allocated to them are fixed for all blocks. In *threshold coding*, however, the locations and the bit allocation can be adapted to the characteristics of the block. For this reason, this method is employed by most video coding standards. In threshold coding, the retained coefficients are selected on the basis of maximum magnitude. Thus, only those coefficients whose

---

[17]For example, if a 256×256 frame was divided into 256 blocks, each of 16×16 pels, then a direct implementation of the 2-D DCT will require: blocks $\times N^4 = 256 \times 16^4 = 16,777,216$ multiplications. If, however, the same frame was divided into 4096 blocks, each of 4×4 pels, then $4096 \times 4^4 = 1,048,576$ multiplications will be required.

magnitudes are above a threshold are retained. In practice, the thresholding and the following quantization operations are combined in one operation using a uniform threshold quantizer as was described in Section 2.5.4 (see Figure 2.5 and Equations (2.8) and (2.10)). In this case, a quantization matrix is used to define the quantizer step size, θ, for each coefficient in the block. A typical quantization matrix is given in Figure 2.10(b). Note that low-frequency coefficients (toward top-left corner) are more finely quantized (i.e., quantized with a smaller step size) because of two reasons. First, the DCT tends to concentrate most of the energy in low frequencies. Second, the HVS is more sensitive to variations in low frequencies. Since in threshold coding the locations of the retained coefficients vary from block to block, those locations need to be encoded. A commonly used strategy is to zigzag scan the transform coefficients, as illustrated in Figure 2.10(c), in an attempt to produce long runs of zeros, and then RLE is used to encode the resulting array.

Compared to predictive coding, transform coding provides higher compression with less sensitivity to errors and less dependence on the input data statistics. Its higher computational complexity and storage requirements have been offset by advances in integrated circuit technology. One disadvantage, however, is that when compression factors are pushed to the limit, three types of artefacts start to occur: (i) "graininess" due to coarse quantization of some coefficients, (ii) "blurring" due to the truncation of high-frequency coefficients, and (iii) "blocking artefacts," which refer to artificial discontinuities appearing at the borders of neighboring blocks due to independent processing of each block. Since blocking artefacts are the most disturbing, a number of methods have been proposed to reduce them. Examples are overlapping blocks at the encoder [34], the use of the lapped orthogonal transform (LOT) [35], and postprocessing using filtering and image restoration techniques [36].

2.6.3 Subband Coding

As already mentioned, rate-distortion theory can provide insights into the design of efficient coders. For example, in Ref. 37 it is shown that the mathematical form of the rate-distortion function suggests that an efficient coder splits the original signal into spectral components of infinitesimal bandwidth and encodes these spectral components independently. This is the basic idea behind subband coding. Subband coding was first introduced by Crochiere *et al.* in 1976 in the context of speech coding [38] and was applied to image coding by Woods and O'Neil in 1986 [39]. In subband coding the input image is passed through a set of bandpass filters to create a set of bandpass images, or *subbands*. Since a bandpass image has a reduced bandwidth compared to the original image, it can be downsampled (subsampled or decimated). This

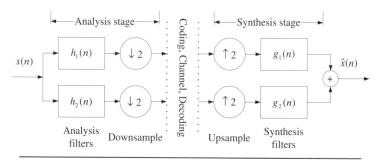

$h_1(n), g_1(n)$: low-pass filters $h_2(n), g_2(n)$: high-pass filters

Figure 2.11: A 1-D, two-band subband coding system

process of filtering and downsampling is called the *analysis* stage. The sub-bands are then quantized and coded independently. At the decoder, the de-coded subbands are upsampled (interpolated), filtered, and added together to reconstruct the image. This is knows as the *synthesis* stage. Note that sub-band decomposition does not lead to any compression in itself, since the total number of samples in the subbands is equal to the number of samples in the original image (this is known as *critical decimation*). The power of this method resides in the fact that each subband can be coded efficiently accord-ing to its statistics and visual importance. A block diagram of a basic 1-D, two-band subband coding system is presented in Figure 2.11.

Ideally, the frequency responses of the low-pass and high-pass filters should be nonoverlapping but contiguous and have unity gain over their bandwidths. In practice, however, filters are not ideal and their responses must be over-lapped to avoid frequency gaps. The problem with overlapping is that aliasing is introduced when the subbands are downsampled. A family of filters that cir-cumvent this problem is the *quadrature mirror filter* (QMF). In the QMF, the filters are designed in such a way that the aliasing introduced by the analysis stage is exactly cancelled by the synthesis stage.

The 1-D decomposition can easily be extended to 2-D using separable filters. In this case, 1-D filters can be applied first in one dimension and then in the other dimension. Using a 1-D two-band decomposition in each direction results in four subbands: horizontal low/vertical low (LL), horizon-tal low/vertical high (LH), horizontal high/vertical low (HL), and horizon-tal high/vertical high (HH), as illustrated in Figure 2.12(a). This four-band decomposition can be continued by repetitively splitting all subbands (uni-form decomposition) or just the LL subband (nonuniform decomposition). A three-stage nonuniform decomposition is illustrated in Figure 2.12(b).

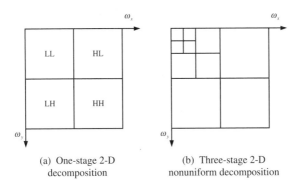

(a) One-stage 2-D
decomposition

(b) Three-stage 2-D
nonuniform decomposition

Figure 2.12: Two-dimensional subband decomposition

Note that nonuniform decomposition results in a multiresolution pyramidal representation of the image. A commonly used technique for nonuniform decomposition is the *discrete wavelet transform* (DWT). The DWT is a transform that has the ability to operate at various scales and resolution levels. Having used the DWT for decomposition, various methods can be used to encode the resulting subbands. One of the most efficient methods is the embedded zero-tree wavelet (EZW) algorithm proposed by Shapiro [40]. This algorithm assumes that if a coefficient at a low-frequency band is zero, it is highly likely that all the coefficients at the same spatial location at all higher frequencies will also be zero and, thus, can be discarded. The EZW algorithm encodes the most important information first and then progressively encodes less important refinement information. This results in an embedded bitstream that can support a range of bit rates by simple truncation. Further refinements to the EZW algorithm have been proposed by Said and Pearlman [41, 42]. In particular, the set partitioning in hierarchical trees (SPIHT) algorithm [42] has become the choice of most practical implementations.

One advantage of subband coding systems is that, unlike transform systems, they do not suffer from blocking artefacts at very low bit rates. In addition, they fit naturally with progressive and multiresolution transmission. One disadvantage, however, is that at very low bit rates, ringing artefacts start to occur around high-contrast edges. This is due to the Gibbs phenomenon of linear filters. To avoid this artefact, subband decomposition using nonlinear filters has been proposed [43, 44].

2.6.4   Vector Quantization

Vector quantization (VQ) is a block-based spatial-domain method that has become very popular since the early 1980s. In VQ, the input image data is

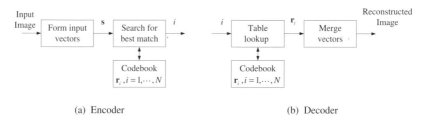

(a) Encoder   (b) Decoder

Figure 2.13: A vector quantization system

first decomposed into k-dimensional input *vectors*. Those input vectors can be generated in a number of different ways; they can refer to the pel values themselves or to some appropriate transformation of them. For example, a $k = M \times M$ block of pels can be ordered to form a k-dimensional input vector $\mathbf{s} = [s_1, \ldots, s_k]^T$. In VQ, the k-dimensional space $\mathscr{R}^k$ is divided into N regions, or *cells*, R_i. Any input vector that falls into cell R_i is represented by a representative *codevector* $\mathbf{r}_i = [r_1, \ldots, r_k]^T$. The set of codevectors $\mathscr{C} = \{\mathbf{r}_1, \ldots, \mathbf{r}_N\}$ is called the *codebook*. Thus, the function of the encoder is to *search* for the codevector $\mathbf{r}_i$ that best matches the input vector $\mathbf{s}$ according to some distortion measure $d(\mathbf{s}, \mathbf{r}_i)$. The index i of this codevector is then transmitted to the decoder using at most $I = \log_2 N$ bits. At the decoder, this index is used to *lookup* the codevector from an identical codebook. A block diagram of a vector quantization system is illustrated in Figure 2.13.

Compression in VQ is achieved by using a codebook with relatively few codevectors compared to the number of possible input vectors. The resulting bit rate of a VQ is given by I/k bits/pel. In theory, as $k \to \infty$, the performance of VQ approaches the rate-distortion bound. However, large values of k make codebook storage and searching impractical. Values of $k = 4 \times 4$ and $N = 1024$ are typical in practical systems.

A very important problem in VQ is the codebook design. A commonly used approach for solving this problem is the *Linde-Buzo-Gray* (LBG) algorithm [45], which is a generalization of the Lloyd-Max algorithm for scalar quantization. The LBG algorithm computes a codebook with a locally minimum average distortion for a given training set and given codebook size. Entropy-constrained vector quantization (ECVQ) [46] extends the LBG algorithm for codebook design under an entropy constraint. Another important problem is the codebook search. A full search is usually impractical, and a number of fast-search algorithms have been proposed, e.g., Ref. 47.

There are many variants of VQ [29]. Examples include adaptive VQ, classified VQ, tree-structured VQ, product VQ (including gain/shape VQ, mean/residual VQ, and interpolative/residual VQ), pyramid VQ, and finite-state VQ.

Theoretically, VQ is more efficient than scalar quantization for both correlated and uncorrelated data [48]. Thus, the scalar quantizer in predictive, transform, and subband coders can be replaced with a vector quantizer.

Vector quantization has a performance that rivals that of transform coding. Although the decoder complexity is negligible (a lookup table), the high complexity of the encoder and the high storage requirements of the method still limit its use in practice. Like transform coding, VQ suffers from blocking artefacts at very low bit rates.

2.6.5 Second-Generation Coding

The coding methods discussed so far are generally known as *waveform coding methods*. They operate on pels or blocks of pels based on statistical image models. This classical view of the image coding problem has three main disadvantages. First, it puts more emphasis on the codeword assignment (using information and coding theory) rather than on the extraction of representative messages. Because the encoded messages (pels or blocks) are poorly representative in the first place, a saturation in compression is eventually reached no matter how good is the codeword assignment. Second, the encoded entities (pels or blocks) are consequences of the technical constraints in transforming scenes into digital data, rather than being real entities. Finally, it does not place enough emphasis on exploiting the properties of the HVS. Efforts to utilize models of the HVS and to use more representative coding entities (real objects) led to a new class of coding methods known as the *second-generation coding methods* [49].

Second-generation methods can be grouped into two classes: *local-operator-based techniques* and *contour/texture-oriented techniques*. Local-operator-based techniques include *pyramidal coding* and *anisotropic nonstationary predictive coding*, whereas the contour/texture-oriented techniques include *directional decomposition coding* and *segmented coding*. Two commonly used segmented coding methods are *region-growing* and *split-and-merge*. For a detailed discussion of second-generation methods, the reader is referred to Refs. 49, 50, 51.

Second-generation methods provide higher compression than waveform coding methods at the same reconstruction quality. They also do not suffer from blocking and blurring artefacts at very low bit rates. However, the extraction of real objects is both difficult and computationally complex. In addition, such methods suffer from unnatural contouring effects, which can make the details seem artificial.

2.6.6   Other Coding Methods

There are many other intraframe coding techniques. Examples are block-truncation coding, fractal coding, quad-tree and recursive coding, multiresolution coding, and neural-network-based coding. A detailed (or even a brief) discussion of such techniques is beyond the scope of this book, and the interested reader is referred to Ref. 52.

2.7   Interframe Coding

As already discussed, video is a time sequence of still images or frames. Thus, a naive approach to video coding would be to employ any of the still-image (or intraframe) coding methods discussed in Section 2.6 on a frame-by-frame basis. However, the compression that can be achieved by this approach is limited because it does not exploit the high temporal correlation between the frames of a video sequence. *Interframe coding* refers to video coding techniques that achieve compression by reducing this temporal redundancy. For this reason, such methods are also known as *temporal redundancy reduction* techniques. Note that interframe coding may not be appropriate for some applications. For example, it would be necessary to decode the complete interframe coded sequence before being able to randomly access individual frames. Thus, a combined approach is normally used in which a number of frames are intraframe coded (I-frames) at specific intervals within the sequence and the other frames are interframe coded (predicted or P-frames) with reference to those anchor frames. In fact, some systems switch between interframe and intraframe within the same frame.

2.7.1   Three-Dimensional Coding

The simplest way to extend intraframe image coding methods to interframe video coding is to consider 3-D waveform coding. For example, in 3-D transform coding based on the DCT, the video is first divided into blocks of $M \times N \times K$ pels (M, N, K denote the horizontal, vertical, and temporal dimensions, respectively). A 3-D DCT is then applied to each block, followed by quantization and symbol encoding, as illustrated in Figure 2.14. A 3-D coding method has the advantage that it does not require the computationally intensive process of motion estimation (as will be discussed in Section 2.7.2). However, it requires K frame memories both at the encoder and decoder to buffer the frames. In addition to this storage requirement, the buffering process limits the use of this method in real-time applications because encoding/decoding cannot begin until all of the next K frames are available. In practical systems, K is typically set to 2–4 frames.

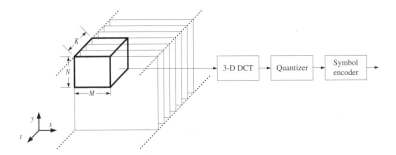

Figure 2.14: A 3-D transform coding system

2.7.2 Motion-Compensated Coding

One of the earliest approaches to interframe coding was *conditional replenishment* (CR) [53]. In this method, the input frame is divided into "changed" and "unchanged" regions with respect to a previously decoded reference frame, and the addresses of this segmentation are coded. Unchanged regions need not be coded because they can simply be copied from the reference frame, whereas changed regions need to be coded. One way of coding them is to use one of the intraframe coding methods discussed in Section 2.6. However, a more efficient approach is to predictively code them with respect to the corresponding regions in the reference frame. In this case, the coded prediction error signal is called the *frame difference* (FD) and the process is known as *frame differencing*.

An improved performance can be obtained by improving the prediction of changed regions. This can be achieved using motion estimation and compensation. Changes between frames are mainly due to the movement of objects. Using a model of the motion of objects between frames, the encoder estimates the motion that occurred between the reference frame and the current frame. This process is called *motion estimation* (ME). The encoder then uses this motion model and information to move the contents of the reference frame to provide a better prediction of the current frame. This process is known as *motion compensation* (MC), and the prediction so produced is called the *motion-compensated prediction* (MCP) or the *displaced-frame* (DF). In this case, the coded prediction error signal is called the *displaced-frame difference* (DFD). A block diagram of a motion-compensated coding system is illustrated in Figure 2.15. This is the most commonly used interframe coding method.

The reference frame employed for ME can occur temporally before or after the current frame. The two cases are known as *forward prediction* and *backward prediction*, respectively. In *bidirectional prediction*, however, two reference frames (one each for forward and backward prediction) are employed

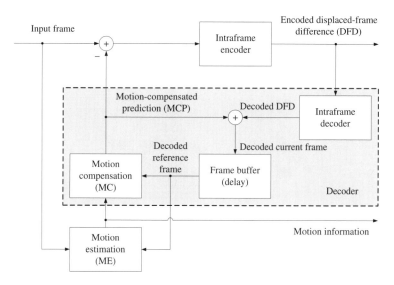

Figure 2.15: Motion-compensated coding system

and the two predictions are interpolated (the resulting predicted frame is called B-frame). The most commonly used ME method is the *block-matching motion estimation* (BMME) algorithm [54]. In this algorithm, the current frame is first divided into blocks. The motion of each block is then estimated by searching for the *best-match* block in the reference frame according to some distortion measure. This search is usually restricted to a *search window* centered around the corresponding block in the reference frame. The motion of the current block is then represented by a *motion vector*, which is the displacement between the block and its best-match block in the reference frame. The process of BMME is illustrated in Figure 2.16. Note that this algorithm is based on a *translational* model of the motion of objects between frames. It also assumes that all pels within a block undergo the same translational movement. There are many other ME methods, but BMME is normally preferred due to its simplicity and good compromise between prediction quality and motion overhead [55]. A more detailed discussion of BMME and other ME methods is deferred to Chapter 4.

As illustrated in Figure 2.15, the DFD signal can be coded using any of the intraframe coding methods discussed in Section 2.6. However, the most commonly used method is transform coding, in particular block-based DCT transform coding. This combination of block-matching motion-compensated prediction and block-based DCT coding of the prediction error has proved to be the most successful class of video coding methods. Today, most video

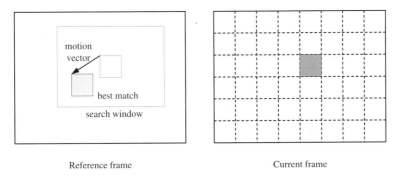

Figure 2.16: Block-matching motion estimation

coding standards are based on this so-called *hybrid* MC-DPCM/DCT coding method.

2.7.3   Model-Based Coding

At very low bit rates (below 64 kbits/s), the quality produced by conventional motion-compensated coding methods may be unacceptable for some applications [10]. In particular, at such bit rates, decoded frames using MC-DPCM/ DCT generally suffer from blocking artefacts. This is mainly due to the translational block-based motion model. This has initiated research efforts into new motion-compensated methods based on more realistic structural motion models. Such methods are referred to as *model-based coding* methods.

Model-based coding is also known as *analysis-synthesis coding*, because it is characterized by two main processes: *analysis* and *synthesis*. Both processes usually make extensive use of sophisticated computer vision and computer graphics tools. At the encoder, the image sequence is initially segmented into a number of objects. Each object is then *analyzed* to decide its location, shape, and texture. The encoder then uses this analysis data to deform a general model to *synthesize* an approximation of the object. The same analysis data is also transmitted to the decoder to synthesize a similar approximation. When the object starts moving, tracking techniques are used, at the encoder, to estimate the associated animation data, which is then transmitted to the decoder to animate the same object. While animation data is sufficient for low quality reproduction at low bit rates, residual data can also be transmitted to achieve higher quality reproduction, but at the expense of a higher bit rate. Thus, once the whole scene is synthesized, only a few animation parameters and possibly some texture information need to be encoded. Hence, model-based coding offers a potential saving in bit rate, which makes it attractive for very-low-bit-rate applications.

Model-based coding methods can be broadly classified as *object-based* or *knowledge-based*. Object-based coding methods deal with unknown (arbitrary) objects, whereas knowledge-based coding methods assume *a priori* knowledge of the objects being modeled (e.g., a 3-D wireframe face model is usually employed for head-and-shoulders sequences typical of videophone applications). Knowledge-based coding methods are generally successful in tracking the global motion of the object (e.g., rotation and translation of the head), but suffer from errors in estimating local motion (e.g., the movement of the eyes, lips, and so on). *Semantic-based coding* is a subset of knowledge-based coding methods that models local motion using a set of action units (e.g., a combination of facial action units can lead to a given facial expression).

Despite their good performance at very low bit rates, model-based coding methods have their problems. For example, at lower bit rates, the analysis and modeling processes become more complex and the model needs to be more object specific. In addition, the analysis and tracking methods usually require some degree of human intervention or some *a priori* assumptions about the nature of tracked objects. Another problem is that, in some cases, severe or sustained failure of tracking or modeling may occur, leading to an increase in the bit rate or a deterioration in the video quality. However, continuous research efforts in this area are addressing such problems. For example, switched model-based coders, with a fallback mode to conventional coding, have been proposed to solve the problem of model or tracking failure [56]. For a good review of model-based coding, the reader is referred to Ref. 57.

Chapter 3

Video Coding: Standards

3.1 Overview

This chapter gives a brief introduction to video coding standards. Section 3.2 highlights the need for video coding standards. Section 3.3 outlines the chronological development of video coding standards, highlighting their main techniques and targeted applications. The chapter then gives two examples of the state-of-the-art video coding standards: Section 3.4 concentrates on H.263 (and its recent extensions: H.263+ and H.263++), whereas Section 3.5 describes MPEG-4.

3.2 The Need for Video Coding Standards

For the past 25 years or so, the efficient coding of image and video signals has been the subject of considerable research. Over the years, the field has matured and has become a key enabling technology for a wide range of applications spanning a wide range of industries. This has moved the field from being a purely academic research area to become a highly commercial business. This increased commercial interest has ignited the efforts of international standardization of image and video coding. International standards enable image and video material from different sources and industries to be processed on different hardware platforms, to be stored on different storage devices, and to be transmitted on different communication networks. This interoperability opens a huge market for video equipment and at the same time gives consumers a wide range of services. International standards also allow for large scale production at considerably reduced costs.

3.3   Chronological Development

Video coding standardization activities started in the early 1980s. The activities were initiated by the *International Telegraph and Telephone Consultative Committee* (CCITT), which is currently known as the *International Telecommunications Union — Telecommunication Standardization Sector* (ITU-T). This was later followed by CCIR (currently ITU-R), the *International Organization for Standardization* (ISO), and the *International Electrotechnical Commission* (IEC). This has resulted in a number of standards, some of which are discussed here.

3.3.1   H.120

The first video coding international standardization activity was carried out by Study Group (SG) XV of CCITT during its study period 1980–1984. In 1984 it issued Recommendation H.120 in its first version, and in 1988 it issued the second version [58]. The standard was targeted for videoconferencing applications at the digital primary rates of 1.544 Mbits/s and 2.048 Mbits/s. The standard had three parts: Part 1 for 625/50 regional use at 2 Mbits/s, Part 3 for 525/60 regional use at 1.5 Mbits/s, and Part 2 for international use (both 525/60 and 625/50 at 1.5 Mbits/s). Parts 1 and 2 use CR with intrafield DPCM for changed regions, whereas Part 3 uses intrafield prediction, background prediction,[1] and motion compensated interfield prediction. This difference in coding techniques between the different parts was one of the reasons why H.120 never became a commercial success.

3.3.2   H.261

At the end of 1984, CCITT/SG XV agreed to define a standard targeted for videophone and videoconferencing applications at ISDN subprimary rates (≤ 2 Mbits/s). Initially, it was thought that there would be two different algorithms efficient at 64 kbits/s or higher and 384 kbits/s or higher, respectively. It was found, however, that a single algorithm could cover all these rates. Thus, H.261 was drafted in 1989 to provide audiovisual services at $p \times 64$ kbits/s ($p = 1 \ldots 30$). This draft became an international standard in 1991 and was later revised in 1993 [59]. H.261 was the first widespread commercial success. In fact, its adopted techniques of hybrid MC-DPCM/DCT (16×16 macroblocks for MC and 8×8 blocks for DCT), SKIP/INTER/

---

[1]None of the later standards have included a background prediction mode, although sprite coding in MPEG-4 can be considered a form of background prediction.

INTRA mode switching on a macroblock level, zigzag scanning, RLE, scalar quantization, and VLC entropy coding have become key elements in most video coding standards.

3.3.3 CCIR-721

In parallel to the standardization activities of CCITT, CCIR started standardization of video coding for contribution-quality TV signals. Recommendation 721 [60] was issued in 1990. Its main target was the transmission of component coded digital TV signals for contribution-quality applications at bit rates near 140 Mbits/s. The recommendation used a simple form of intrafield DPCM to achieve a low implementation complexity and a high degree of random access (which is important for video postprocessing in studios).

3.3.4 CCIR-723

CCIR Recommendation 723 [61] was issued in 1992. Its main target was the coding of component digital TV signals for contribution-quality applications in the range 34–45 Mbits/s. The recommendation employs a hybrid MC-DPCM/DCT with one intrafield mode and two interfield modes (with and without MC). Both the CCIR-721 and the CCIR-723 recommendations are not generic. In contrast to other standards, they fully specify both the encoder and the decoder.

3.3.5 MPEG-1

In 1988, the *Moving Picture Experts Group* (MPEG) was created under Subcommittee (SC) 2 of ISO (ISO/SC2). The group is now Working Group (WG) 11 of SC29 under the Joint Technical Committee (JTC) 1 of ISO/IEC. Thus, its official denotation is ISO/IEC JTC1/SC29/WG11. The main aim of the group was to develop a video coding standard for digital media storage applications at up to 1.5 Mbits/s. In 1991 the group drafted its ISO/IEC 11172 (MPEG-1) standard [62], which became an international standard in 1992. The MPEG-1 video algorithm is very similar to the H.261 algorithm but with some advanced techniques, like bidirectional prediction and half-pel MC[2]. The standard also provides for some specific storage requirements, like random access and fast forward/reverse searches. Although the standard was developed mainly for storage applications, it was designed to be generic. Thus, it was

---

[2]Half-pel MC was proposed during the development of the H.261 but was thought to be too complex at that time.

designed as a toolbox, where the user can decide which tools to use for the particular application. In addition, the standard defines only the decoder and the bitstream syntax. This allows a large degree of freedom for manufacturers to propose their own optimized encoders. This generic design and large degree of freedom have contributed to the success of MPEG-1. It has been used in a wide range of applications, from interactive systems on CD-ROM to the delivery of video over telecommunication networks.

3.3.6   MPEG-2

In 1990, ISO/IEC JTC1/SC29/WG11 started studies on a new standard for applications not covered by MPEG-1. In particular, the new standard was intended to provide video quality not lower than NTSC/PAL and up to CCIR-601 quality at rates around 10 Mbits/s. This standardization activity was nicknamed MPEG-2 because it was seen as phase 2 of the work started in MPEG-1. In 1992, ITU-T/SG 15 joined this standardization effort to develop video coding for *Asynchronous Transfer Mode* (ATM) networks. In 1993, it was realized that the scope of MPEG-2 could be enlarged to suit coding of HDTV. This made an initially planned MPEG-3 for HDTV superfluous. In 1994, the ISO/IEC 13818 (MPEG-2) standard (ITU-T Recommendation H.262) was drafted [63], and later in the year it was accepted as an international standard. Like MPEG-1, the MPEG-2 standard is generic and flexible. In fact, MPEG-2 can be thought of as a superset of, and as such was designed to be backward compatible with, MPEG-1. There are many additional features provided by MPEG-2 over MPEG-1, including the support for interlaced video and scalability. Since implementation of the full MPEG-2 syntax may not be practical for most applications, MPEG-2 has introduced the concepts of "profiles," describing functionalities, and "levels," describing resolutions, to provide subset conformance levels. MPEG-2 has had even more success than MPEG-1, with applications in the areas of cable TV, networked ATM services, and satellite and terrestrial TV broadcasting.

3.3.7   H.263

The increasing demand for digital video communications over the public switched telephone network (PSTN) and mobile networks initiated a new standardization effort by ITU-T/SG 15. The aim was to develop a video coding standard for low-bit-rate applications below 64 kbits/s. The result of this effort was ITU-T Recommendation H.263 [64], which was completed in 1995 and approved in 1996. Although H.263 was based on the coding structure of H.261, it provides a significant improvement in performance. Side-by-side comparisons indicate that H.263 provides the same subjective quality as H.261 but with less than half the bit rate [65]. This performance improvement is due to optimized coding techniques as well as advanced optional coding modes. Some of the new features of H.263 compared to H.261 are the support for more picture formats, half-pel MC, a 3-D (LAST-RUN-LEVEL) RLE instead of 2-D (RUN-LEVEL), more optimized VLC tables, optional extra headers to increase error resilience, advanced 2-D median predictor for motion vector coding, more optimized macroblock addressing and quantization adaptation, optional extended-range unrestricted motion vectors that can point outside frames, optional arithmetic coding, optional advanced prediction with overlapped motion compensation and four motion vectors per macroblock, and optional bidirectional prediction. H.263 is described in more detail in Section 3.4.

3.3.8   H.263+

Technically, H.263+ is version 2 of the H.263 standard [66]. This version was developed by ITU-T/SG16/Q15 Advanced Video Experts Group (previously under ITU-T/SG15), with technical content completed in 1997 and approved in 1998. The H.263+ standard added 12 new optional features to H.263. These new features support custom picture size and clock frequency, improve compression efficiency, allow scalability, enhance error resilience over wireless and packet-based networks, provide supplemental display and external usage capabilities, and ensure backward compatibility. H.263+ is described in more detail in Section 3.4.

3.3.9   MPEG-4

In 1993, the ISO/IEC JTC1/SC29/WG11 MPEG group initiated a new standardization activity called MPEG-4. The target was the very-low-bit range and the aim was to achieve higher compression efficiency than could be achieved by existing conventional techniques. In 1994, it was realized that too few improvements could be achieved over the H.263 and H.263+ compression results to justify a new standard. Thus, the group decided to broaden the objectives of the MPEG-4 effort and started an in-depth analysis of the trends within the audiovisual world. Particular attention was given to the convergence of the three traditionally separate industries of communications, computing, and TV/film/entertainment. This study concluded that MPEG-4 should support functionalities that would be useful in future applications but were not supported or not well supported by the available standards. Eight main new or improved functionalities were identified and then clustered in three classes:

content-based interactivity (content-based multimedia data-access tools, content-based manipulation and bitstream editing, hybrid natural and synthetic data coding, improved temporal random access), *compression* (improved coding efficiency, coding of multiple concurrent data streams), and *universal access* (robustness in error-prone environments, content-based scalability). Version 1 of the MPEG-4 standard was approved in October 1998. A second version was approved in December 1999 to add new functionalities and improve others. The MPEG-4 standard is officially known as ISO/IEC 14496 and is titled "Generic coding of audiovisual objects" [67]. This title describes two important properties of the MPEG-4 standard. The first property is that it is a generic standard. It is designed to cover a wide range of bit rates (typically, 5 kbits/s to 10 Mbits/s), picture formats (progressive and interlaced), resolutions (SQCIF to beyond TV), frame rates (still images to high frame rates), communication networks (wired or wireless), input material (natural or synthesized), etc. The second property is that it uses an object-based representation model, where a scene is represented, coded, and manipulated as individual audiovisual objects. This particular property (i.e., being object-based) sets MPEG-4 apart from earlier block-based standards. Thus, in addition to conventional block-based MC-DPCM/DCT techniques, MPEG-4 adopts more recent object-based techniques like second-generation coding techniques (Section 2.6.5) and model-based coding techniques (Section 2.7.3). MPEG-4 is described in more detail in Section 3.5.

3.3.10   H.263++

Technically, H.263++ is version 3 of the H.263 standard [68]. This version was developed by ITU-T/SG16/Q15, with technical content completed and approved late in the year 2000. The H.263++ standard added some more features to H.263 and H.263+. These new features improve coding efficiency, enhance error resilience, provide additional supplemental display and external usage capabilities, and define profiles and levels. H.263++ is described in more detail in Section 3.4.

3.3.11   H.26L

This is a project of ITU-T/SG16/Q15. The H.26L project is planned to be a new-generation video coding standard with improved efficiency, error resilience, and streaming support. It is scheduled for completion in 2002.

In addition to the standard documents themselves, interested readers are referred to some excellent reviews and tutorials available in the literature [69, 70, 65, 71–75, 11, 13, 15].

3.4 The H.263 Standard

3.4.1 Introduction

The H.263 recommendation specifies a coded representation that can be used for compressing the moving picture component of audiovisual services at low bit rates. The recommendation fully specifies the decoder and the bitstream syntax but does not explicitly specify the encoder. As already mentioned, this gives manufacturers a large degree of freedom to propose their own optimized encoders, as long as the output bitstream conforms to the standard decodable syntax. However, during the standardization process, a software-based codec (encoder-decoder) called the *test model* is developed to study the core elements of the standard. For example, version 5 of the *test model near-term* (TMN) is described in Ref. 76.

3.4.2 Source Format

The standard supports all five members of the CIF family described in Section 2.4.4 and Table 2.1. As a minimum requirement, all decoders shall be able to operate with SQCIF and QCIF. Encoders, on the other hand, shall be able to operate with either SQCIF or QCIF and are not obliged to be able to operate with both.

3.4.3 Video Source Coding Algorithm

The generalized form of the source coder is illustrated in Figure 3.1. It is a hybrid of interpicture prediction to utilize temporal redundancy and transform coding of the error signal to reduce spatial redundancy.

3.4.3.1 Picture Coding Structure

The input video consists of a sequence of pictures (or frames). Each picture is divided into *groups of blocks* (GOBs). A GOB consists of $k \times 16$ lines, depending on the picture format ($k = 1$ for SQCIF, QCIF, and CIF, $k = 2$ for 4CIF, and $k = 4$ for 16CIF). For example, there are 9 GOBs in a QCIF picture. Each GOB is divided into *macroblocks* (MBs). A macroblock consists of 16×16 samples of Y' and the spatially corresponding 8×8 samples of C_B' and C_R'. If we define a *block* as 8×8 samples of Y', C_B', or C_R', then a macroblock consists of 6 blocks: 4 luma blocks and the 2 spatially corresponding chroma blocks. Figure 3.2 illustrates the H.263 picture structure for a QCIF frame. As shown, GOBs are coded from top to bottom in increasing number. Within each GOB, the MBs are coded from left to right (and from top to bottom if the GOB contains more than one row of MBs) in increasing number. Within

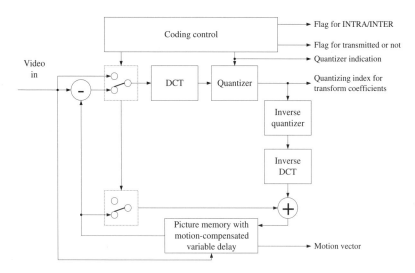

Figure 3.1: H.263 video encoder

each MB, the Y' blocks are first coded in the order shown (left to right and top to bottom), followed by the C'_B block and then the C'_R block.

3.4.3.2   Coding Modes

The coding mode in which interpicture prediction is applied is called the INTER mode. Prediction can optionally be augmented by motion compensation. If no prediction is applied, then the coding mode is called INTRA. The coding mode (INTRA/INTER) can be signaled at the picture level (resulting in I-pictures/P-pictures) or at the macroblock level in P-pictures. In PB-frames (discussed later) the B-pictures are always coded in INTER mode. The mode selection method is not defined by the standard. However, to control the accumulation of IDCT mismatch[3] error, the standard requires an MB to be coded in INTRA mode at least once every 132 times when coefficients are transmitted for this MB in P-pictures. In the INTER mode, a flag is used to indicate whether an MB is transmitted or not (conditional replenishment). This is sometimes referred to as the SKIP mode. Again, the method of reaching a

---

[3]The *inverse discrete cosine transform* (IDCT) is a common block between the encoder and the decoder. Differences in implementation between the encoder's IDCT and the decoder's IDCT cause mismatches between the reconstructed pictures at both ends. This is called the *IDCT mismatch error*. This mismatch accumulates due to interpicture prediction and can be stopped by forced INTRA updating.

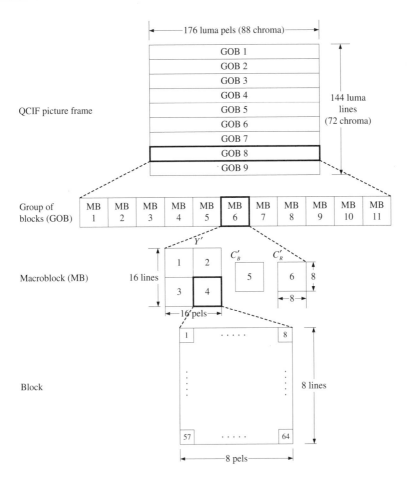

Figure 3.2: H.263 picture structure for a QCIF frame

decision to transmit an MB or not is not part of the standard. The different flags are encoded within the picture and MB headers.

3.4.3.3 Motion Estimation and Compensation

Without options, the encoder estimates one motion vector per MB. Both horizontal and vertical components of the vector have integer or half-integer values and are restricted to the range $[-16, +15.5]$. A positive value of the horizontal or vertical component means that the prediction is made from pels in the reference picture that are spatially to the right or below the pels being predicted, respectively. Motion vectors are restricted such that all pels referenced by them are within the reference picture area. The standard does not explicitly specify an ME method. However, this MB-based structure implicitly supports block-based approaches and in particular the BMME algorithm.

3.4.3.4 Motion Vector Coding

The estimated motion vector $\mathbf{MV} = (\mathrm{MVx}, \mathrm{MVy})$ is predictively coded. This means that the motion vector difference $\mathbf{MVD} = \mathbf{MV} - \mathbf{MVP}$ is encoded instead of the motion vector itself. The motion vector predictor $\mathbf{MVP}$ is the median of three candidate predictors, which are the motion vectors of three surrounding macroblocks, as illustrated in Figure 3.3(a). The two components of the motion vector difference are then entropy coded using a standard VLC table. MVDx is encoded first, followed by MVDy.

3.4.3.5 Forward Transform

The forward DCT transform is applied either to the pel values, in the case of an INTRA MB, or to the DFD values, in the case of an INTER MB. In both cases, the DCT is applied on a block (8×8) basis. This results in six blocks of transform coefficients for each MB. The standard does not specify

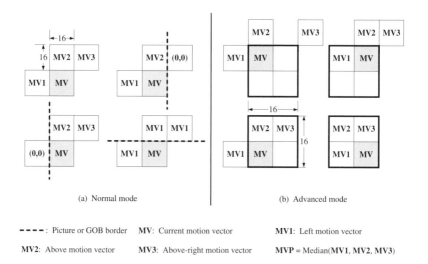

(a) Normal mode   (b) Advanced mode

- - - - : Picture or GOB border   **MV**: Current motion vector   **MV1**: Left motion vector

MV2: Above motion vector   **MV3**: Above-right motion vector   **MVP** = Median(**MV1**, **MV2**, **MV3**)

Figure 3.3: H.263 motion vector prediction

the method of implementing the forward DCT. Threshold coding, discussed in Section 2.6.2, is used to allocate bits to the transform coefficients, as will be discussed next.

3.4.3.6 Quantization

The six DC coefficients of an INTRA MB are quantized using a uniform scalar quantizer with a step size of $\theta = 8$ and no dead zone (this corresponds to Figure 2.5(a) and Equation (2.8)). All other coefficients are quantized using a uniform scalar quantizer with a step size of $\theta = 2 \times QP$ and a central dead zone around zero (this corresponds to Figure 2.5(b) and Equation (2.10)). There are 31 possible quantization parameters, $QP = 1 \ldots 31$. However, the quantization parameter is kept fixed for all coefficients within an MB. A high QP leads to higher compression but worse quality, whereas a low QP leads to better quality but less compression. The method to select a QP is not part of the standard. A change of QP to any of the 31 permissible values can be signaled in the picture or GOB headers. In the MB header, however, this change is limited to a maximum of ± 2. Again, the method to decide this change is not defined in the standard.

3.4.3.7 Quantized Coefficients Coding

A quantized INTRA DC coefficient is encoded using a standard 8-bit FLC table. Other quantized coefficients are first zigzag scanned, as described in Section 2.6.2 and Figure 2.10(c). The reordered coefficients are then encoded using 3-D RLE. Thus, the reordered coefficients are converted to an intermediate set of symbols or EVENTS of the form (LAST, RUN, LEVEL), where LAST is an indication of whether this is the last nonzero coefficient in the block or not, RUN is the number of successive zeros preceding the coded coefficient, and LEVEL is the nonzero value of the coded coefficient. The most commonly occurring EVENTs are coded using a standard VLC table, whereas the remaining EVENTs are coded using a concatenation of four standard FLC codewords for ESCAPE, LAST, RUN and LEVEL.

3.4.3.8 Coding Control

The coding control block is responsible for varying several parameters to control the rate or the quality of the coded video. Examples are the INTER/INTRA mode decision at the picture or MB level, the update pattern of the forced INTRA refresh, the TRANSMIT/SKIP decision at the MB level, and the QP and its change at the picture, GOB, or MB level. Such functions are not defined in the standard.

3.4.4 Decoding Process

3.4.4.1 Motion Vector Decoding

For each TRANSMITTED INTER MB, the decoder calculates the same motion vector predictor **MVP** used at the encoder and adds it to the decoded motion vector difference **MVD** to obtain the decoded motion vector **MV**. The motion vector of a SKIPPED INTER MB is set to **0**.

3.4.4.2 Motion Compensation

The decoded motion vector is used to compensate the four Y' blocks in the MB. Motion vectors for both C_R' and C_B' blocks are derived by dividing the component values of the decoded motion vector by 2. The resulting quarter-pel resolution components are modified toward the nearest half-pel resolution (both 0.25 and 0.75 are rounded to 0.5). If motion compensation requires accessing half-pel positions, then bilinear interpolation is used to calculate the pel values at those positions.

3.4.4.3 Inverse Quantization

As already discussed, quantization is achieved by dividing the transform coefficient by a quantization step size and rounding the result (refer to Equations (2.8) and (2.10)). Inverse[4] quantization is the process of reconstructing an approximation of the original coefficient by multiplying the quantized coefficient by the same step size (refer to Equations (2.9) and (2.11)). The reconstructed coefficients are then clipped to the range $[-2048, +2047]$ and inverse zigzag scanned to put them in an 8×8 block.

3.4.4.4 Inverse Transform

The reconstructed block of coefficients is processed by a separable 2-D 8×8 inverse DCT. The arithmetic procedures for computing the inverse DCT are not defined by the standard, but should meet a defined error tolerance.

3.4.4.5 Reconstruction of Blocks

For INTRA blocks, the reconstructed block is equal to the result of the inverse DCT. For INTER blocks, the reconstructed block is formed by summing the motion-compensated prediction and the result of the inverse DCT. The reconstructed values are clipped to the range $[0, 255]$.

---

[4]It should be emphasised that the term *inverse* here does not mean that quantization is a reversible process. Quantization is irreversible since rounding leads to loss of information.

3.4.5 Optional Coding Modes

There are four optional coding modes that can be signaled at the picture level. These modes are defined in annexes to the standard and are briefly described next.

3.4.5.1 Unrestricted Motion Vector Mode (Annex D)

In this mode, motion vectors are allowed to point outside the reference picture area. When a pel pointed to by a motion vector is outside the reference picture area, an edge pel is used instead. This edge pel is found by limiting the motion vector to the last full-pel position inside the reference picture area. Limitation of the motion vector is performed on a pel basis and separately for each component of the motion vector. In this mode also, the range for motion vector components is extended to $[-31.5, +31.5]$, with the restriction that if the predictor is in the range $[-15.5, +16]$, then only values that are within a range of $[-16, +15.5]$ around the predictor can be reached. If, however, the predictor is outside $[-15.5, +16]$, then all values within the range $[-31.5, +31.5]$ with the same sign as the predictor can be reached. Allowing motion vectors to point outside the reference picture area improves prediction along picture edges in the case of camera or background movement. This is particularly useful for small picture formats (where border MBs represent a high percentage of the picture area). The extended motion vector range allows better prediction for large picture formats and a high amount of movement.

3.4.5.2 Syntax-Based Arithmetic Coding Mode (Annex E)

In this mode, all VLC Huffman coding/decoding operations of H.263 are replaced with arithmetic coding/decoding operations. As already discussed in Section 2.5.5, arithmetic coding removes the restriction of representing each symbol by an integral number of bits, achieving more coding efficiency but at the expense of more computational complexity.

3.4.5.3 Advanced Prediction Mode (Annex F)

This optional mode includes two advanced prediction techniques: the use of four motion vectors per MB, and the use of *overlapped motion compensation* (OMC). In addition, this mode allows motion vectors to point outside the reference picture area. If this mode is used in combination with the unrestricted motion vector mode, then the motion vectors will also have an extended range. If the mode is used in combination with the PB-frames mode, then OMC is only used for P-pictures, not for B-pictures.

In this mode, the encoder makes a decision (which is not defined by the standard) whether to transmit one motion vector or four motion vectors per

MB. If one motion vector is transmitted (as in normal mode), then the decoder replicates it to four motion vectors. If four motion vectors are to be transmitted, then the motion vector prediction process is modified as illustrated in Figure 3.3(b). Motion vectors for chroma blocks are derived by calculating the sum of the four luma vectors and then dividing by 8. The resulting values of 1/16-pel resolution are modified toward the nearest 1/2-pel values (0, 1/16, and 2/16 are modified to 0; 14/16 and 15/16 are modified to 1; and all other values are modified to 1/2). This technique improves prediction if the MB contains different moving objects.

In OMC, each pel in an 8×8 luma prediction block is predicted as a weighted sum of three prediction values. To obtain the three prediction values, three motion vectors are used: the motion vector of the current luma block, and two out of four *remote* motion vectors. The four remote motion vectors are the motion vectors of the luma blocks to the left of, to the right of, above, and below the current luma block. The position of the pel within the block decides which two remote vectors to use. For example, all pels in the top-left quadrant of the block use the two remote vectors of the blocks above and to the left of the current luma block. The weight given to each of the three predictions also changes with pel position within the block. The weights are defined in three standard matrices. The weights for a remote prediction are designed to increase as the pel position moves away from the center of the block toward the corresponding remote block. This ensures a smooth transition at the borders of the block, which results in a visible reduction of blocking artefacts. If a remote MB was not coded, then the corresponding vector is set to zero. If a remote MB does not exist (out of the picture) or was INTRA coded, then the corresponding vector is set to the vector of the current MB. In PB-frames mode, however, INTRA MBs have motion vectors, and those are used as remote vectors. For chroma blocks, no overlapping is performed.

3.4.5.4 PB-Frames Mode (Annex G)

In this mode, two pictures are encoded as one unit called a PB-frame. Thus, a PB-frame consists of one P-picture that is predicted from the previous decoded P-picture (forward prediction) and one B-picture that is predicted from both the previous decoded P-picture and the P-picture currently decoded in the same PB-frame (bidirectional prediction). In a PB-frame, an MB consists of 12 blocks: the 6 blocks of the P-picture, followed by the 6 blocks of the B-picture. In this mode, an INTRA coding mode can also be used where P-blocks are INTRA coded and B-blocks are INTER coded with prediction as for an INTER block. In this case, motion vector data is included with the INTRA-coded P-blocks but are used for predicting B-blocks.

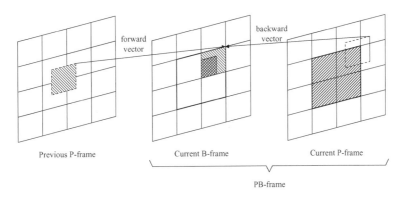

Figure 3.4: Prediction in PB-frames mode

For prediction of a B-block, both forward, MV_F, and backward, MV_B, motion vectors are needed. Those are not transmitted but are derived at the decoder by scaling the corresponding P-block motion vector, MV, using the temporal resolutions of the P- and B-pictures with respect to the previous P-picture. The derived motion vectors can be optionally enhanced using a transmitted delta vector $MVDB$. The forward (or backward) motion vectors for chroma blocks are derived by summing the corresponding luma forward (or backward) motion vectors, dividing by 8, and then rounding to the nearest half-pel resolution. To be able to predict a B-macroblock, the corresponding P-macroblock is first reconstructed. For pels of a B-block where MV_B points outside the reconstructed P-macroblock, forward prediction using MV_F and the previous decoded P-picture is used. However, for pels of the same B-block where MV_B points inside the reconstructed P-macroblock, bidirectional prediction is used. In this case, the prediction is the average (with truncation) of the forward prediction, using MV_F and the previous decoded P-picture, and the backward prediction, using MV_B and the reconstructed P-macroblock. This process is illustrated in Figure 3.4. With this mode, the frame rate can be increased without a significant increase in bit rate.

3.4.6   H.263, Version 2 (H.263+)

Version 2 of the H.263 standard is informally known as H.263+. This version adds a number of optional feature enhancements to version 1. In the process of adding these new features, the precise definition and requirements of the original version 1 syntax and semantics were not changed. In fact, version 2 is backward compatible with version 1. The additional optional feature set can be summarized in terms of the new types of pictures, a modified unrestricted motion vector mode, and 12 new optional modes (annexes I–T). This is briefly described in what follows.

3.4.6.1   New types of pictures

Version 2 defines three new types of pictures:

1. *Scalability pictures*: Three types of scalability pictures were added, one that provides temporal scalability and two that provide SNR or spatial scalability:

    (a) B: a picture having two reference pictures, one of which temporally precedes the B picture and one of which is temporally subsequent to the B picture.

    (b) EI: a picture having a temporally simultaneous reference picture.

    (c) EP: a picture having two reference pictures, one of which temporally precedes the EP picture and one of which is temporally simultaneous.

    These pictures are described in more detail in Section 3.4.6.9 in the discussion of the new optional scalability mode (annex O).

2. *Improved PB-frames*: Recent investigations have indicated that the current PB-frames utilized by version 1 are not sufficiently robust for continual use. Encoders implementing the PB-frames mode are limited to use only bidirectional prediction. In some situations, this results in a lack of usefulness of the PB-frames mode. An improved, more robust type of PB-frames has been added to enable heavier, higher-performance use of the PB-frames mode. This is described in more detail in Section 3.4.6.7 in the discussion of the new optional improved PB-frames mode (annex M).

3. *Custom source formats*: As already discussed, version 1 allows only five video source formats (CIF family) with defined picture size, picture shape, and picture clock frequency. Version 2, however, allows a wide range of optional custom source formats in order to make the standard apply to a much wider class of video scenes and applications, such as resizable computer window-based displays, high refresh rates, and wide-format viewing screens.

3.4.6.2   Modified Unrestricted Motion Vector Mode (modified Annex D)

The optional unrestricted motion vector mode (annex D) of version 1 has been modified in version 2. Version 2 defines a new data field called PLUSPTYPE. When using the unrestricted motion vector mode, if PLUSPTYPE is present in the picture header, then the following modifications apply:

1. The motion vector range no longer depends on the motion vector prediction value. There are two cases here:

   (a) If the UUI data field in the picture header is set to "1," the motion vector range depends on the picture format. For standardized picture formats up to CIF the range is $[-32, 31.5]$, for those up to 4CIF the range is $[-64, 63.5]$, for those up to 16CIF the range is $[-128, 127.5]$, and for even larger custom picture formats the range is $[-256, 255.5]$. In addition, the horizontal and vertical motion vector ranges may be different for custom picture formats.

   (b) If, however, the UUI data field is set to "01," the motion vectors are not limited except by their distance to the coded area border, as explained by the following restriction rule: the motion vector values are restricted such that no element of the 16×16 (or 8×8) region that is selected shall have a horizontal or vertical distance more than 15 pels outside the coded picture area.

2. A new VLC table is employed to encode the motion vector differences. This table has the following properties:

   (a) The codes are single-valued. In other words, each codeword corresponds to a single motion vector difference value. This is in contrast to the double-valued VLC codes of version 1, where each codeword can represent one of two possible motion vector differences. Double-valued codes were not popular due to their high implementation cost and the limitations on their extendibility.

   (b) The table employs *reversible variable-length coding* (RVLC) codewords. Such codewords can be decoded in both the forward and backward directions. As discussed in Chapter 9, the use of RVLCs can increase the error resilience of video bitstreams. In addition, RVLCs are easier to implement because they can easily be generated and decoded using a simple state machine.

3.4.6.3 Advanced INTRA Coding Mode (Annex I)

This optional mode significantly improves the compression performance when coding INTRA macroblocks. The mode is applied both to INTRA macroblocks within INTRA-pictures and to INTRA macroblocks within INTER-pictures. The improved compression performance of this mode is achieved as follows:

1. INTRA blocks are predicted from their neighboring INTRA blocks. Block prediction always uses data from the same luma or chroma

component. There are three options for prediction:

(a) DC only: where the DC coefficient is predicted as the average of the corresponding coefficients from the block above and the block to the left.

(b) Vertical DC and AC: where the DC coefficient and the first row of AC coefficients are vertically predicted from the corresponding coefficients from the block above.

(c) Horizontal DC and AC: where the DC coefficient and the first column of AC coefficients are horizontally predicted from the corresponding coefficients from the block to the left.

Special cases are defined for situations in which the neighboring blocks are not INTRA coded or are not in the same video picture segment. The option that gives the best prediction for the whole macroblock is chosen.

2. The quantization of INTRA coefficients is modified. INTRA DC coefficients are quantized using a varying quantization step size, unlike the fixed quantization step size of 8 utilized when this mode is not in use. In addition, the quantization of all INTRA coefficients is performed without a dead-zone.

3. The scanning of DCT coefficients is adapted to the prediction method of the INTRA macroblock. For macroblocks predicted using the DC-only option, the normal zigzag scanning is utilized; for macroblocks predicted using the vertical DC and AC option, a new alternate horizontal scanning pattern (Figure 3.5(a)) is utilized; whereas for macroblocks predicted

0	1	2	3	10	11	12	13
4	5	8	9	17	16	15	14
6	7	19	18	26	27	28	29
20	21	24	25	30	31	32	33
22	23	34	35	42	43	44	45
36	37	40	41	46	47	48	49
38	39	50	51	56	57	58	59
52	53	54	55	60	61	62	63

(a) Alternate horizontal scan

0	4	6	20	22	36	38	52
1	5	7	21	23	37	39	53
2	8	19	24	34	40	50	54
3	9	18	25	35	41	51	55
10	17	26	30	42	46	56	60
11	16	27	31	43	47	57	61
12	15	28	32	44	48	58	62
13	14	29	33	45	49	59	63

(b) Alternate vertical scan

Figure 3.5: Alternate scans for the advanced INTRA mode of H.263+

using the horizontal DC and AC option, a new alternate vertical scanning pattern (Figure 3.5(b)) is utilized.

4. The quantized INTRA coefficients are encoded using a new VLC table optimized for the global statistics of INTRA macroblocks.

3.4.6.4 Deblocking Filter Mode (Annex J)

In this optional mode, a filter is applied, both at the encoder and at the decoder, across the boundaries of luma and chroma 8×8 blocks of reconstructed pictures before storing them in the picture memory. In other words, the filter affects the picture that is used for the prediction of subsequent pictures and thus lies within the motion prediction loop.

The deblocking filter operates using a set of four pel values either on a horizontal or on a vertical line of the reconstructed picture. Two of the four pels belong to one block, whereas the other two belong to a neighboring block. The weights of the filter's coefficients depend on the quantizer step size, where stronger coefficients are used for a coarser quantizer, and vice versa. No filtering is performed across a picture edge. Similarly, when the *Independent Segment Decoding* (ISD) mode is in use, no filtering is performed across slice edges (when the Slice Structured mode is in use) or across the top boundary of GOBs having GOB headers present (when the Slice Structured mode is not in use). When this mode is used together with the Improved PB-frames mode, the backward prediction of the B-macroblock is based on the reconstructed P-macroblock before the deblocking edge filter operations. The mode applies only for the P-, I-, EP-, or EI-pictures or the P-picture part of an Improved PB-frame. Possible filtering of B-pictures or the B-picture part of an Improved PB-frame is not a matter for standardization.

In addition to the filtering operation, this mode allows the use of four motion vectors per macroblock and also the use of unrestricted motion vectors. This mode improves the prediction quality and significantly reduces blocking artefacts.

3.4.6.5 Slice Structured Mode (Annex K)

In this optional mode, a *slice* layer is employed instead of the normal GOB layer. This mode is used to provide enhanced error resilience, to make the bitstream more amenable to use with packet-based networks, and to minimize video delay. A slice layer allows a flexible partitioning of the picture into segments containing a variable number of macroblocks. It also allows more control over the shape of segments. In addition, a slice structure provides more flexibility in the transmission order. This is in contrast with a GOB layer, which only allows partitioning into fixed-size, fixed-shape segments with fixed transmission order.

In order to facilitate optimal usage in a number of environments, this mode contains two submodes:

1. *The Rectangular Slice (RS) submode*: When RS is in use, the slice occupies a rectangular region of width specified in units of macroblocks and contains a number of macroblocks in scanning order within the rectangular region. When RS is not in use, the slice contains a number of macroblocks in scanning order within the picture as a whole.

2. *The Arbitrary Slice Ordering (ASO) submode*: when ASO is in use, the slices may appear in any order within the bitstream. When ASO is not in use, the slices must be sent in scanning order.

A slice video picture segment starts at a macroblock boundary in the picture and contains a number of macroblocks. Different slices within the same picture shall not overlap with each other, and every macroblock shall belong to one and only one slice.

A *slice* is defined as a slice header followed by consecutive macroblocks in scanning order. In order to allow slice header locations within the bitstream to act as resynchronization points for bit error and packet loss recovery and in order to allow out-of-order slice decoding within a picture, slice boundaries are treated differently than simple macroblock boundaries. Thus, no data dependencies can cross the slice boundaries within the current picture. An exception to this is the Deblocking Filter mode, which, when in use without the Independent Segment Decoding mode, filters across the boundaries of the blocks in the picture.

3.4.6.6   Supplemental Enhancement Information Mode (Annex L)

In this mode, additional supplemental information can be included in the bit-· stream to signal an enhanced display capability or to provide information for external usage. The supplemental information may be present in the bitstream even though the decoder may not be capable of providing the enhanced capability to use it or even to properly interpret it. In this case, the decoder can simply discard the supplemental information. The mode can be used to signal the following capabilities:

1. *Picture Freeze*: The mode can be used to signal that the contents of the entire prior-displayed picture, or a specified rectangular part of it, shall be kept unchanged. The mode can also be used to explicitly signal a picture freeze release.

2. *Picture Freeze with Resizing*: The mode can be used to signal that the contents of a specified rectangular area of the prior-displayed picture

should be resized to fit into a smaller part of the displayed video picture, which should then be kept unchanged.

3. *Picture Snapshot*: The mode can be used to signal that the current picture, or a specified rectangular part of it, is labeled for external use as a still-image snapshot of the video content.

4. *Video Time Segment*: The mode can be used to signal the beginning and the end of a specified subsequence of video data to be used externally.

5. *Progressive Refinement Segment*: The mode can be used to signal the beginning and the end of a specified subsequence of video data. Rather than being a continually moving scene, this subsequence of video includes a start picture followed by a sequence of zero or more pictures to refine its quality.

6. *Chroma-Keying Information*: The mode can be used to indicate that the *chroma-keying* technique is used to represent *transparent* and *semitransparent* pels in the decoded video pictures. When being presented on the display, transparent pels are not displayed. Instead, a background picture is revealed that is either a prior reference picture or an externally controlled picture. Semitransparent pels are displayed by blending the pel value in the current picture with the corresponding value in the background picture.

3.4.6.7 Improved PB-Frames Mode (Annex M)

This mode represents an improvement compared to the original PB-frames optional mode (annex G). The main difference between the two modes is that the original PB-frames mode can utilize only bidirectional prediction to predict the B part in a PB-frame, whereas the improved PB-frames mode can utilize forward, backward, or bidirectional prediction.

The bidirectional prediction method is the same as in the original PB-frames mode, except that in this case no delta vector is transmitted. In the forward-prediction method, a B macroblock is predicted from the previously decoded P-picture and a forward motion vector is transmitted. In the backward-prediction method, a B-macroblock is predicted from the corresponding P-macroblock currently decoded in the same PB-frame, and therefore no backward motion vector needs to be transmitted.

This mode significantly improves coding efficiency in situations in which downscaled P-vectors (utilized in the original PB-frames mode) are not good candidates for B-prediction. In particular, the backward prediction is useful when there is a scene cut between the previous P-frame and the current PB-frame. In general, it is advisable to use the Improved PB-frames mode instead of the original PB-frames mode.

3.4.6.8   Reference Picture Selection Mode (Annex N)

In normal operation, a picture is temporally predicted from the most recently decoded picture. The reference picture section (RPS) mode, however, allows temporal prediction from pictures other than the most recently decoded one. Thus, in this mode, both the encoder and the decoder use more than one picture memory. As discussed in Chapter 6, this method belongs to a class of motion estimation and compensation techniques called *multiple-reference motion-compensated prediction*. The information to signal which picture is selected for prediction is included by the encoder in the encoded bitstream. However, the strategy used by the encoder to select this picture is not subject for standardization.

This mode can be used to improve the performance of video communication over error-prone channels. In normal operation, if part of the reference picture is lost due, for example, to a transmission error, then this error will propagate to and severely degrade the quality of future pictures. In this mode, however, the encoder may switch to another reference picture to suppress the temporal error propagation due to interframe coding.

In order to utilize this mode, the encoder needs to have some knowledge about the conditions of the channel and the outcome of the decoding process (e.g., which parts of the reference picture have been decoded in error). One way to achieve this is to utilize a backward (feedback) channel. This mode has two back-channel mode switches that define whether a backward channel is used and what kind of messages are returned on that backward channel from the decoder. Together, the two switches define four basic methods of operation: NEITHER (no backward messages), ACK (acknowledgment messages only), NACK (negative acknowledgment messages only), and ACK+NACK (both acknowledgment and negative acknowledgment messages). There are also two methods of operation in terms of the channel for backward channel messages. The first method is the Separate Logical Channel mode, where back-channel data is delivered through a separate logical channel in the multiplex layer of the system, whereas the second method is the VideoMux mode, where back-channel data for received video is delivered within the forward video data of a video stream of encoded data.

3.4.6.9   Temporal, SNR, and Spatial Scalability Mode (Annex O)

Scalability implies that a bitstream is composed of a *base layer* and one or more associated *enhancement layers*. The base layer is separately decodable. The enhancement layers can be decoded in conjunction with the base layer to increase perceived quality by either increasing the picture rate (temporal scalability), increasing the picture SNR quality (SNR scalability), or increasing the picture resolution (spatial scalability). This mode has support for three types of scalability: temporal, SNR, and spatial scalability, as detailed next. This mode can be helpful when used over heterogenous networks with varying bandwidth capacity and also in conjunction with error correction schemes.

a. **Temporal scalability:** *Temporal scalability* refers to enhancement information used to increase the picture quality by increasing the picture display rate. Temporal scalability is achieved by employing bidirectionally predicted pictures, or B-pictures. B-pictures can be predicted from a previous and/or a subsequent reconstructed picture in the reference layer (the layer used for prediction). B-pictures in this mode differ from the B-picture part of a PB- (or an Improved PB-) frame in that they are separate entities in the bitstream. In other words, they are not syntactically intermixed with a subsequent P-picture. It should be emphasised that B-pictures should not be used as reference pictures for the prediction of any other picture. This is particularly important to allow for B-pictures to be discarded if necessary without adversely affecting any subsequent pictures, thus providing temporal scalability. Figure 3.6(a) illustrates temporal scalability using B-pictures. It should be pointed out that the location of B-pictures in the bitstream is in a data-dependence order rather than in a temporal order. For example, in the case shown in Figure 3.6(a) the bitstream order of the encoded pictures is I_1, P_3, B_2, P_5, B_4,.... . There is no limit to the number of B-pictures that may be inserted between pairs of reference pictures in the reference layer. In this mode, motion vectors are allowed to extend beyond the picture boundaries of B-pictures.

b. **SNR scalability:** *SNR scalability* refers to enhancement information used to increase the picture quality without increasing picture resolution. The process of compression usually introduces artefacts and distortions. As a result, the difference between a reconstructed picture and its original in the encoder is almost always a nonzero-valued picture. Normally, this coding error picture is lost at the encoder and never recovered. With SNR scalability, however, these coding-error pictures can be encoded and sent to the decoder. At the decoder, such coding-error pictures can be used to increase the signal-to-noise ratio of the decoded picture, and hence the term SNR scalability. Figure 3.6(b) illustrates SNR scalability. If the enhancement-layer picture is predicted only from a simultaneous lower-layer reference picture, then the enhancement-layer picture is referred to as an EI-picture. If, however, the enhancement-layer picture is bidirectionally predicted using both a prior enhancement-layer picture and a temporally simultaneous lower-layer reference picture, then the enhancement-layer picture is referred to as an EP-picture. The picture in the reference

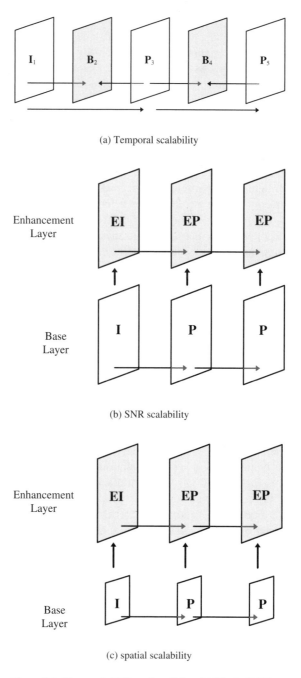

(a) Temporal scalability

(b) SNR scalability

(c) spatial scalability

Figure 3.6: Temporal, SNR, and spatial scalability in H.263+

layer that is used for upward prediction of an EI- or EP-picture may be an I-picture, a P-picture, or the P part of a PB- or Improved PB-frame. Thus, an EI-picture in an enhancement layer may have a P-picture as its lower-layer reference picture, and an EP-picture may have an I-picture as its lower-layer enhancement picture. For both EI- and EP-pictures, the prediction from the lower reference layer uses no motion vectors. However, EP-pictures use motion vectors for the prediction from their prior reference picture in the same layer.

c. **Spatial scalability:** *Spatial scalability* refers to enhancement information used to increase the picture quality by increasing picture resolution either horizontally, vertically, or both. Spatial scalability is very similar to SNR scalability. The only difference is that before the picture in the reference layer is used to predict the picture in the enhancement layer, it is interpolated by a factor of 2 either horizontally or vertically (1-D spatial scalability) or both horizontally and vertically (2-D spatial scalability). The interpolation filters for this operation are defined by the standard. Spatial scalability is illustrated in Figure 3.6(c).

d. **Multilayer scalability:** It is possible not only for B-pictures to be temporally inserted between pictures of types I, P, PB, and Improved PB, but also between pictures of types EI and EP (whether these consist of SNR or spatial-enhancement pictures). It is also possible to have more than one SNR or spatial-enhancement layers in conjunction with a base layer. Thus a multilayer scalable bitstream can be a combination of SNR layers, spatial layers, and B-pictures.

3.4.6.10   Reference Picture Resampling Mode (Annex P)

In this mode, a resampling operation can be applied to the previously decoded picture in order to generate a new *warped* picture for use as reference for predicting the currently encoded picture. For example, if the previous reference picture and the current picture are of different source formats, then this mode can be used to resample the previous picture to match the source format of the current picture. Another example is to use this mode to warp the previous reference picture to compensate for global motion. Warping and warping-based motion estimation methods are discussed in Chapter 5.

3.4.6.11   Reduced-Resolution Update Mode (Annex Q)

This mode allows the encoder to send information encoded at a low resolution to update a higher-resolution reference picture and produce a final picture at the higher resolution. This mode is particularly useful when encoding a highly active scene, and allows an encoder to increase the picture rate at which moving parts of a scene can be represented while maintaining a higher-resolution representation in more static areas of the scene.

The syntax of the bitstream in this mode is identical to the syntax for coding without the mode, but the semantics, or interpretation of the bitstream, is somewhat different. In this mode, the portion of the picture covered by a macroblock is twice as wide and twice as high. Thus, there is approximately one-quarter the number of macroblocks as there would be without this mode. Motion vector data also refers to blocks of twice the normal height and width, or 32×32 and 16×16 instead of the normal 16×16 and 8×8. For example, the decoder receives and decodes a 16×16 DFD block at the reduced resolution. The decoder then upsamples this block to 32×32 at the higher resolution. The decoder then upsamples the received motion vector by a factor of 2 and uses it to produce a 32×32 prediction from the reference picture. The DFD block and the prediction block are then added to produce a 32×32 block at the higher resolution.

3.4.6.12 Independent Segment Decoding Mode (Annex R)

This mode allows a picture to be constructed without any data dependencies that cross video picture segments. Thus, this mode provides error robustness by preventing the propagation of erroneous data across the boundaries of video picture segments.

In this mode, a video picture segment can be a slice, a GOB or multi-GOBs with nonempty GOB headers, or a complete picture. When this mode is in use, the video picture segment boundaries are treated as picture boundaries. In other words, each video picture segment is decoded with complete independence from all other video picture segments, and is also independent of all data outside the corresponding video picture segment in the reference picture(s).

For example, motion vectors of blocks outside the current video picture segment cannot be used when calculating the current motion vector predictor. Similarly, motion vectors of blocks outside the current video picture segment cannot be used as remote motion vectors for overlapped block-motion compensation when the Advanced Prediction mode is in use. In addition, no motion vectors are allowed to reference areas outside the corresponding video picture segment in the reference picture(s).

3.4.6.13 Alternative INTER VLC Mode (Annex S)

This mode improves the efficiency of encoding some INTER macroblocks by allowing a VLC table originally designed for INTRA macroblocks to be used for some INTER macroblocks. The INTRA VLC table used in the advanced INTRA coding mode (annex I) is designed to efficiently encode INTRA blocks. Thus, it is optimized for coding blocks with many large-valued coefficients and small runs of zeros. There are cases, however, where the statistics of INTER blocks can approximate the statistics of INTRA blocks. This is particularly possible when significant changes are evident in the picture or when small quantizer step sizes are employed. In such cases, it can become more efficient to encode INTER blocks using the INTRA VLC table.

In this mode, the encoder would normally choose to use the INTRA VLC table for coding an INTER block only when the use of this table results in fewer bits than the use of the INTER VLC table. This use of the alternative INTRA VLC table, however, is subject to the condition that the decoder would be able to detect which of the two tables was used for encoding. Thus, the alternative INTRA VLC table can be used, subject to the condition that decoding using the INTER VLC table would result in runs of zeros so long as to indicate the presence of more than 64 coefficients in the block.

3.4.6.14 Modified Quantization Mode (Annex T)

In this mode, the quantizer operation is modified. In particular, this mode includes the following four key features:

1. In normal mode, the change of the quantization parameter at the macroblock level is limited to a maximum of ± 2. This mode, however, improves the bit-rate control ability by allowing the quantization parameter to be changed at the macroblock level to any of its 31 permissible values.

2. In normal mode, the chroma quantizer step size is the same as that for luma. This mode, however, improves the fidelity of chroma by specifying a smaller quantizer step size for chroma than that for luma.

3. The true value of a DCT coefficient prior to quantization can be as high as 2040. Thus, when the quantization parameter is less than 8, the quantized DCT coefficients can be outside the range $[-127, +127]$ permissible in the normal mode. Such coefficients are clipped to the permissible range before being encoded. This mode, however, extends the range of representable quantized DCT coefficient values to allow the representation of any possible true coefficient value to within the accuracy allowed by the quantizer step size.

4. In this mode certain restrictions are placed on the encoded DCT coefficient values to improve the detectability of errors and to minimize decoding complexity.

Kossentini *et al.* [71, 72] provide an excellent overview of H.263+ and evaluate the performance of the modes individually and in different combinations.

3.4.7   H.263, Version 3 (H.263++)

Version 3 of the H.263 standard is informally known as H.263++. This version adds a number of optional feature enhancements to versions 1 and 2.

3.4.7.1   Enhanced Reference Picture Selection Mode (Annex U)

The *enhanced reference picture selection* (ERPS) mode is an enhancement to the RPS mode (annex N) of H.263+. In addition to enhancing error resilience, this mode provides benefits in terms of coding efficiency.

As with the RPS mode, the ERPS mode extends the motion estimation and compensation processes to use more than one reference picture. In the ERPS mode, however, enhanced performance is achieved by allowing reference picture selection on the macroblock, rather than the picture, level. Thus, in this case, each motion vector is extended by a picture reference parameter that is used to address a macroblock or block prediction region in any of the multiple reference pictures.

The ERPS mode also includes a submode for improving the coding efficiency of B-pictures. In this submode, encoders can use more than one reference picture for both forward and backward prediction of B-pictures.

Another submode of ERPS is provided to reduce memory requirements. In this submode, each reference picture is partitioned into smaller rectangular units called *subpictures*. The encoder can then indicate to the decoder that specific subpicture areas of specific reference pictures will not be used as a reference for the prediction of subsequent pictures. This allows the memory allocated in the decoder for storing these areas to be used to store data from other reference pictures.

3.4.7.2   Data Partitioned Slice Mode (Annex V)

In this mode, data is arranged in a video picture segment as defined in the independent segment decoding mode (annex R) of H.263+. The contents of this segment are rearranged such that the header information for *all* the MBs in the segment are encoded and transmitted *together*, followed by the motion vectors for all the MBs in the segment and then by the DCT coefficients for all the MBs in the segment. The segment header uses the same syntax as the slice structured mode (annex K) of H.263+. The header, motion vectors, and DCT partitions are separated by markers. In addition to data partitioning, this mode uses RVLC tables for encoding header and motion information. As will be discussed later, data partitioning and RVLC provide robustness in error-prone environments. Another error-resilience enhancement in this mode is that the motion vector predictor is no longer formed from three neighboring motion vectors. Instead, a new prediction method is used to allow independent motion vector decoding in both the forward and backward directions.

3.4.7.3 Additional Supplemental Enhancement Information (Annex W)

This annex describes additional supplemental enhancement information that adds to the functionality of the supplemental enhancement information mode (annex L) of H.263+. In particular, the following additional information can be added to the bitstream:

1. Indication of the use of a specific fixed-point IDCT.

2. Picture messages, including the message types of:

    (a) Arbitrary binary data.

    (b) Text (arbitrary, copyright, caption, video description, or uniform resource identifier).

    (c) Picture header repetition (current, previous, next with reliable temporal reference, or next with unreliable temporal reference).

    (d) Interlaced field indications (top or bottom).

    (e) Picture number.

    (f) Spare reference picture identification.

3.4.7.4 Profiles and Levels Definitions (Annex X)

With the variety of optional modes available in H.263, it is crucial that several preferred mode combinations for operation be defined so that different terminals will have a high probability of connecting to each other. This annex contains a list of preferred mode combinations, which are structured into "profiles" of support. It also defines some groupings of maximum performance parameters as "levels" of support for these profiles.

The annex defines nine profiles (profile 0 to profile 8). Each profile is defined in terms of a set of features supported by the decoder. For example, the *Baseline Profile* (profile 0) refers to the syntax of H.263 with no optional modes of operation. Another example is *Version 2 Interactive and Streaming Wireless Profile* (profile 3). This profile is defined to provide enhanced coding efficiency performance and enhanced error resilience for delivery to wireless devices within the feature set available in H.263+. This profile of support is composed of the baseline design plus the following modes: advanced INTRA coding mode (Annex I), deblocking filter mode (Annex J), slice structured mode (Annex K), and the modified quantization mode (Annex T).

The annex also defines seven levels (level 10 to level 70) of performance capability for decoder implementation. For example, a decoder supporting the first level, level 10, must include support of QCIF and sub-QCIF resolution decoding, and must be capable of operation with a bit rate up to 64,000 bits per second with a picture decoding rate up to $(15,000)/1001$ pictures per second.

3.5   The MPEG-4 Standard

As already discussed, the formal title "Generic coding of audiovisual objects" given to MPEG-4 describes two important properties of the standard. The first property is that it is a generic standard. It defines tools and algorithms for the coding of natural, synthesis, and hybrid audiovisual objects with a wide range of bit rates, picture formats, transmission media, etc. It is, therefore, very difficult to describe the full functionality of such a generic standard in a volume of this size.[5] Thus, this section will concentrate on MPEG-4 *natural video coding*. In particular, the section will try to highlight the second property of MPEG-4, i.e., being object-based, which sets it apart from other standards.

3.5.1   An Object-Based Representation

MPEG-4 uses an object-based representation model. Thus, a scene is represented, coded, and manipulated as individual *audiovisual objects* (AVOs). This section concentrates on *natural video objects*.

As illustrated in Figure 3.7, an MPEG-4 *video session* (VS) is a collection of one or more *video objects* (VOs). A VO is an entity that a user is allowed to access (e.g., seek and browse) and manipulate (e.g., cut and paste). It can be a simple rectangular frame or it can be an arbitrarily shaped object. A VO can consist of one or more *video object layers* (VOLs). As is discussed later, each VO can be encoded in either a scalable (multiple VOLs) or a nonscalable (single VOL) form. Each VOL consists of an ordered sequence of *video object planes* (VOPs). A VOP is an instance (or a snapshot) of the corresponding VO at a given time. A number of VOPs can, optionally, be grouped together in a *group of video object planes* (GOV). GOVs can provide points in the bitstream where VOPs are encoded independently from each other. This provides random access points within the bitstream.

Figure 3.8 shows a general block diagram of an MPEG-4 codec. The input video is represented using a number of VOs. This object-based representation either already exists (e.g., generated with chroma-key technology) or is

---

[5]To give an indication of how generic the MPEG-4 standard is, the MPEG-4 draft [67] that was used in writing the current section is more than 300 pages.

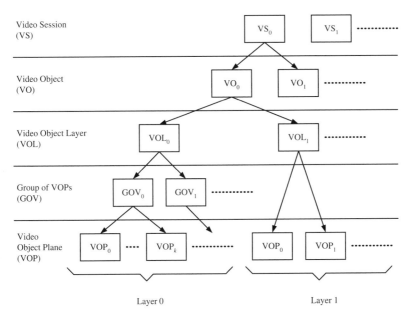

Figure 3.7: MPEG-4 video bitstream structure

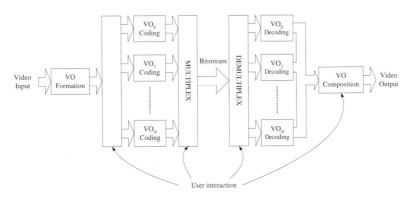

Figure 3.8: An MPEG-4 codec

generated using segmentation techniques. Each VO is encoded individually, and the resulting bitstreams are multiplexed to a single bitstream. At the decoder, the received bitstream is first demultiplexed to the individual bitstreams. Each bitstream is then decoded, and the decoded VOs are composited to reconstruct the output video. As shown, at various points of this encoding-decoding process, users are allowed to interact with (access and/or manipulate) the individual VOs.

As an example, consider a sequence showing a hot-air balloon flying in the sky. In this case, the sequence can be represented using two VOs: the balloon and the sky background. Figure 3.9(a) shows a single frame of this sequence. At this particular instance of time the two VOs are represented by the two VOPs shown in Figures 3.9(b) and 3.9(c). At the encoder, each VOP is encoded individually and the two bitstreams are multiplexed. At the decoder, the received bitstream is demultiplexed to the two individual bitstreams. Each bitstream is then decoded to reconstruct the corresponding VOP. The two VOPs are then put together to reconstruct the transmitted frame. The user can optionally manipulate the decoded VOPs. For example, in Figure 3.9(d) the balloon VOP has been enlarged, rotated, and translated as compared to the original frame.

In addition to composition information (which indicates where and when the VOP is to be displayed), each VOP is encoded in terms of its *shape*,

(a) Balloon in Sky (original)

(b) Sky (background) VOP

(c) Balloon VOP

(d) Decoded and manipulated

Figure 3.9: Object-based representation, coding, and interaction

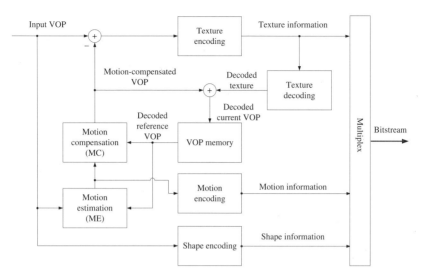

Figure 3.10: An MPEG-4 VOP encoder

motion, and *texture*. This is illustrated in Figure 3.10. As can be seen, an MPEG-4 VOP encoder has three main functionalities: *shape encoding, motion encoding* (along with motion estimation and compensation), and *texture encoding*. Note that the structure of this encoder is very similar to the MC-DPCM structure utilized by H.263 and most other standards. In fact, for most cases, the texture encoder is DCT-based and the structure is very similar to the conventional hybrid MC-DPCM/DCT encoder. The difference here is that the encoded entities can have arbitrary shapes rather than the fixed rectangular frame shape, and therefore additional shape information needs to be encoded and transmitted. Note that this object-based representation can be thought of as a generic representation. When a frame is encoded using a single VOP, this generic representation degenerates into the special case of rectangular frames and an MPEG-4 encoder becomes almost identical to an H.263 encoder. In fact, the MPEG-4 standard provides measures to ensure some level of inter-operability with MPEG-1/2 and H.263.

A VOP is encoded on a macroblock (MB) basis. MPEG-4 supports a $4:2:0$ subsampling format with 4–12 bits/sample. Thus, an MB consists of six 8×8 blocks: four luma blocks and two corresponding chroma blocks. To achieve efficient encoding, the arbitrary shaped VOP is first encapsulated within a bounding box. This bounding box is chosen such that it completely contains the VOP but uses the minimum number of macroblocks. This bounding box is illustrated for the Balloon VOP in Figure 3.11. Within this bounding box,

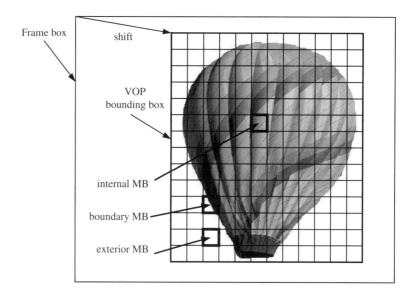

Figure 3.11: The bounding box of the Balloon VOP

there are three types of MBs: internal MBs, boundary MBs, and exterior MBs. An *internal* MB lies completely inside the VOP, whereas a *boundary* MB lies on the contour of the VOP; i.e., parts of it are inside the VOP and the other parts are outside the VOP. An *exterior* MB, on the other hand, lies completely outside the VOP. Note that the shape, size, and location of this bounding box can change from one time instance to another. Thus, the absolute (frame) coordinate system is used to define such bounding boxes.

The following subsections briefly describe the main building blocks of the MPEG-4 VOP encoder.

3.5.2 Shape Coding

In the context of MPEG-4, shape information is referred to as *alpha planes*. There are two types of alpha planes: binary and gray-scale. A *binary alpha plane* defines which pels within the bounding box belong to the video object at a given instant of time. A *gray-scale alpha plane*, on the other hand, is a more general form of alpha planes, for it includes transparency information.

3.5.2.1 Binary Shape Coding

A binary alpha plane is represented by a matrix the same size as the bounding box of the video object. Every element within this matrix can take one of two possible values. If the corresponding pel belongs to the object, then the element is set to 255; otherwise it is set to 0. This matrix is sometimes referred to as a *binary mask* or as a *bitmap*. Figure 3.12 shows the binary alpha plane of the Balloon VOP.

Before encoding, the binary alpha plane is partitioned into 16×16 blocks called *binary alpha blocks* (BABs). A BAB with all elements equal to 0 is called a *transparent* BAB, whereas a BAB with all elements equal to 255 is called an *opaque* BAB. Each BAB is encoded separately. The main tools used for encoding BABs are *context-based arithmetic encoding* (CAE) and motion compensation. There are two variants of the CAE algorithm. One is used with motion compensation and is called InterCAE, whereas the other one is used without motion compensation and is called IntraCAE. There are seven possible modes for encoding a BAB:

1. The BAB is flagged transparent. In this case, no shape coding is necessary. In addition, texture information is not coded for this BAB.

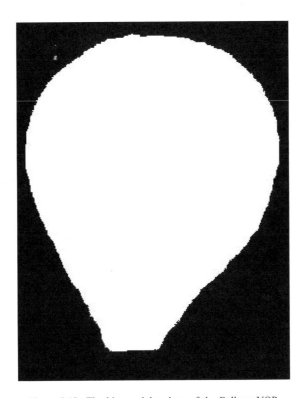

Figure 3.12: The binary alpha plane of the Balloon VOP

2. The BAB is flagged opaque. In this case, no shape coding is necessary, but texture information is coded.

3. The BAB is coded without motion compensation using IntraCAE.

4. The **MVDs** is zero (i.e., **MVs** = **MVPs**) and no block update is necessary.

5. The **MVDs** is zero and the block needs to be updated. In this case, InterCAE is used for coding the block update.

6. The **MVDs** is nonzero and no update is necessary.

7. The **MVDs** is nonzero and the block needs to be updated. In this case, InterCAE is used for coding the block update.

Modes 1 and 2 require no shape coding. For mode 3, shape is encoded using IntraCAE. For modes 4–7, motion estimation and compensation are employed. The motion vector difference for shape (**MVDs**) is the difference between the shape motion vector (**MVs**) and its predictor (**MVPs**). This predictor is estimated from either neighboring shape motion vectors or co-located texture motion vectors. When the mode indicates that no update is required, then the **MVs** is simply used to copy a displaced 16×16 block from the reference binary alpha plane to the current BAB. If, however, the mode indicates that an update is required, then the update is coded using InterCAE.

The CAE is a binary arithmetic coding algorithm where the probability of a symbol is determined from the *context* of the neighboring symbols. First, the arithmetic encoder is initialized. The binary pels (elements) of the BAB are then encoded in raster-scan order using the following steps:

1. Compute a context number based on the templates shown in Figure 3.13. This context number is given by $C = \sum_{k=0}^{N-1} c_k 2^k$, where $c_k = 0$ for a

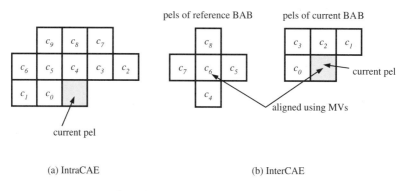

(a) IntraCAE   (b) InterCAE

Figure 3.13: CAE templates in MPEG-4

transparent pel, $c_k = 1$ for an opaque pel, $N = 10$ pels for IntraCAE, and $N = 9$ pels for InterCAE.

2. Determine the probability of the pel being transparent (or opaque) by using the context number to index a table of probabilities defined by the standard.

3. Use the indexed probability to drive an arithmetic encoder for codeword assignment.

When all pels in the BAB have been encoded, the arithmetic encoder is terminated.

3.5.2.2 Gray-Scale Shape Coding

A gray-scale alpha plane has a similar representation to the binary alpha plane, with the difference that elements within the plane can take on a range of values, usually 0 to 255 with 8-bit representation, designating the degree of transparency of the corresponding pel. Gray-scale shape information consists of two parts. The first part is the support information. This is obtained by thresholding the gray-scale alpha plane at 0 (i.e., any value that is not equal to 0 is set to 255). Support information is encoded using the binary shape coding methods described previously. The second part of gray-scale shape information contains the gray-scale values of the alpha plane. This is encoded using methods similar to the texture encoding methods described later in this chapter (Section 3.5.4).

3.5.2.3 Scalable Shape Coding

Besides changing the coding mode of BABs, additional mechanisms are employed for controlling the quality and bit rate of binary shape information. One method is by reducing the resolution of the BAB by a factor of 2 or 4. The resulting 8×8 or 4×4 BAB is encoded using any of the available modes. At the decoder, the reduced-resolution BAB is first decoded and then upsampled. Another method for reducing the binary shape bit rate is by changing the orientation of the BAB. The efficiency of the CAE algorithm can depend on the orientation of the BAB. In some cases, transposing the BAB before coding it can increase coding efficiency. In this case, the decoder decodes the BAB and then transposes it back to its original orientation.

3.5.3 Motion Estimation and Compensation

Motion estimation and compensation methods in MPEG-4 are very similar to those employed by other standards. The main difference is that block-based motion estimation and compensation are adapted to the arbitrary-shape VOP structure of MPEG-4. The standard has three modes for encoding a given VOP: intra-VOP (I-VOP), predicted-VOP (P-VOP), and bidirectionally-predicted-VOP (B-VOP).

Since the shape, size, and location of a VOP can change from one instance to another, the absolute (frame) coordinate system is used for referencing every VOP. Thus, the motion vector for a particular feature inside a VOP refers to the displacement of the feature in absolute coordinates. During motion estimation and compensation, no alignment of VOP bounding boxes at different time instances is performed.

Motion is estimated only for those MBs within the bounding box of the current VOP. If the current MB is an internal MB, then motion is estimated using the usual block-matching method. If, however, the current MB is a boundary MB, then motion is estimated using a modified block-matching method called *polygon matching*. In polygon matching, the distortion measure is calculated using only those pels in the current macroblock that belong to the VOP.

The motion estimation and compensation processes may require accessing pels outside the reference VOP. Padding is used to define the values of such pels. The luma component is padded per 16×16 samples, while the chroma components are padded per 8×8 samples. If the reference MB is a boundary MB, then it is padded using *repetitive padding*. This process starts by horizontal repetitive padding, where each sample at the boundary of a reference VOP is replicated horizontally in the left and/or right direction in order to fill the transparent region of the reference MB. If there are two boundary sample values for filling a sample, the two boundary samples are averaged. The remaining unfilled transparent samples are padded by a similar process as the horizontal repetitive padding but in the vertical direction, i.e., vertical repetitive padding. The remaining MBs within the reference VOP are exterior MBs. Such MBs are filled by *extended padding*. In this method, samples of an exterior MB are filled by replicating the samples at the border of the neighboring boundary MB. If an exterior MB is next to more than one boundary MB, then one of the boundary MBs is chosen according to a priority criterion defined by the standard. The remaining exterior MBs are filled with 128 (for an 8-bit luma component).

Motion vectors are estimated to half-pel accuracy. They are then predictively VLC coded in a similar fashion to the H.263 standard.

Similar to the H.263 standard, MPEG-4 has an advanced prediction mode (four motion vectors per MB and unrestricted motion vectors) and an overlapped motion compensation mode.

3.5.4 Texture Coding

For I-VOPs, texture refers to the luma and chroma values (i.e., the video signal). For motion-compensated VOPs, texture refers to the luma and chroma residual errors remaining after motion compensation (i.e., the DFD signal). The process of texture coding involves the following steps: padding, DCT, quantization, INTRA coefficient prediction, scanning, and variable-length encoding.

3.5.4.1 Padding

Like H.263 and most other video coding standards, MPEG-4 encodes texture information using a block-based 8×8 DCT. In this process, internal MBs are encoded directly, whereas boundary MBs must first be padded. The aim of this padding process is to remove abrupt transitions within the macroblock, thus reducing the number of significant DCT coefficients. Note that during texture coding, exterior MBs are not coded.

For motion-compensated boundary MBs, pels outside the VOP are padded with zero. For INTRA boundary MBs, pels outside the VOP are padded using the following *low-pass extrapolation* (LPE) procedure:

1. Calculate the mean value of the macroblock pels that lie within the VOP. Use this value for padding the macroblock pels that lie outside the VOP.

2. Starting at the top-left corner of the macroblock, proceed in scanning order to the bottom-right corner, replacing each pel $f(x, y)$ that lies outside the VOP with the average value of its four neighbors; i.e., $f(x, y) = (f(x - 1, y) + f(x + 1, y) + (f(x, y - 1) + f(x, y + 1))/4$. The neighboring pels should lie within the VOP; otherwise they are not considered in the averaging process and the equation is modified accordingly.

3.5.4.2 DCT

The internal MBs and the padded boundary MBs are then transformed using a 2-D 8×8 forward DCT.

3.5.4.3 Quantization

The resulting DCT coefficients are quantized using one of two methods. The first method is very similar to H.263 quantization and uses a fixed quantization step size for the whole macroblock. The second method, however, uses one of two default quantization matrices (or scaled versions of them) to modify the quantizer step size depending on the spatial frequency of the coefficient. In MPEG-4, DC coefficients can also be quantized using a nonlinear quantizer.

3.5.4.4 Prediction of INTRA DCT Coefficients

To achieve more efficiency, the quantized coefficients of an INTRA block can be predicted from the colocated coefficients in either the block immediately to the left of or the block immediately above the current block, as shown in Figure 3.14. The direction of prediction is adapted depending on the horizontal and vertical DC gradients of neighboring blocks. Thus, if X is the current INTRA block, $QF_A(0,0)$ is the quantized DC coefficient of block A immediately to the left of the current block, $QF_B(0,0)$ is the quantized DC coefficient of block B above and to the left of X, and $QF_C(0,0)$ is the quantized DC coefficient of block C immediately above X, then the direction of prediction is chosen as follows: If $|QF_A(0,0) - QF_B(0,0)| < |QF_B(0,0) - QF_C(0,0)|$, then predict from block C; otherwise predict from block A.

Having decided the direction of prediction, there are two types of prediction:

1. *DC prediction*: Depending on the direction of prediction, the DC coefficient of the current block X is predicted from the DC coefficient of either block A or block C. For example, when the horizontal direction is chosen, the prediction is given by $PQF_X(0,0) = QF_X(0,0) - QF_A(0,0)$.

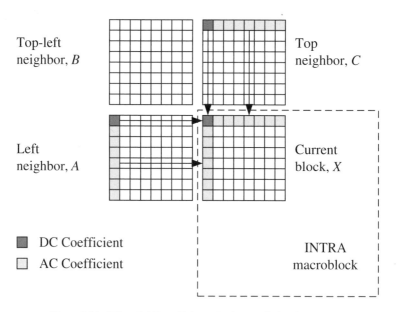

Figure 3.14: DC and AC coefficient adaptive prediction in MPEG-4

2. *AC prediction*: Depending on the direction of prediction, the AC coefficients of the first row of the current block X are predicted from the AC coefficients of the first row of block C, or the AC coefficients of the first column of the current block X are predicted from the AC coefficients of the first column of block A. To compensate for differences in the quantization parameters of adjacent blocks used in AC prediction, the prediction process is modified so that the predictor is scaled by the ratio of the current quantization parameter, QP_X, and the quantization parameter of the predictor block, QP_A or QP_C. For example, when the horizontal direction is chosen, the prediction is given by $PQF_X(0, j) = QF_X(0, j) - \frac{QF_A(0, j)QP_A}{QP_X}$. The use of AC prediction can be enabled/disabled at the macroblock level.

If any of the neighboring blocks are outside of the VOP boundary or the video packet boundary, or if they do not belong to an INTRA coded macroblock, their DC values are assumed to take a value of $2^{bits/pel+2}$ and their AC values are assumed to take a value of 0. DC and AC predictions are performed similarly for the luma and each of the two chroma components.

3.5.4.5 Scanning

To prepare the coefficients for variable-length encoding, a scanning process is used to convert the 2-D matrix of coefficients into a 1-D vector.

There are three possible scanning patterns: zigzag, alternate-vertical, and alternate-horizontal. All non-INTRA blocks use the conventional zigzag scanning pattern. For INTRA blocks, however, the choice of the scanning pattern depends on the prediction process:

1. If AC prediction is *not* employed, then the conventional zigzag scanning pattern is used for all blocks within the macroblock.

2. If, however, AC prediction is employed, then the direction of the DC prediction is used to select a suitable scanning pattern on a block basis, as follows:

   (a) If the DC prediction employs the horizontal direction, then the alternate-vertical scanning pattern is used.

   (b) If, however, the DC prediction employs the vertical direction, then the alternate-horizontal scanning pattern is used.

3.5.4.6 Variable-Length Coding

The differential (predicted) DC coefficients in INTRA macroblocks are encoded using a concatenation of a VLC codeword and a FLC codeword. The possible range of encoded differential DC coefficients is divided into subranges or categories. The VLC codeword indicates to which category the encoded difference belongs, whereas the FLC codeword, then, uniquely identifies the difference within that category. Instead of this special treatment, the INTRA DC coefficients can optionally be encoded using the same INTRA AC VLC table described next. To achieve compatibility with H.263, the INTRA DC coefficients can also optionally be encoded without prediction using an 8-bit FLC codeword.

All other coefficients are encoded using a procedure similar to that of H.263. Thus, the scanned quantized coefficients are converted into an intermediate set of EVENTS of the form (LAST, RUN, LEVEL). The most commonly occurring events are then encoded using standard VLC tables. There are two standard VLC tables: one for INTRA blocks and another for INTER blocks. To achieve compatibility with H.263, the VLC table for INTER blocks can optionally be used for both INTER and INTRA blocks. Less frequent EVENTS are encoded with the help of an ESCAPE codeword.

3.5.5   Still-Texture Coding

The MPEG-4 also supports coding of static textures (or still images). This mode uses subband coding based on the discrete wavelet transform (DWT).

As discussed in Section 2.6.3, the DWT is used in subband coding to apply a nonuniform decomposition (refer to Figure 2.12(b)) to the texture information. This results in a decomposition tree of subbands. The lowest subband (horizontal low/vertical low (LL)) is known as the DC subband, whereas the remaining subbands are known as the AC subbands.

In MPEG-4, the DWT can be either a floating-point or an integer transform, as signaled by the encoder in the bitstream. The encoder can also choose to use a set of default filters or to use its own filters and define them in the bitstream.

The quantized coefficients of the DC subband are encoded using DPCM followed by arithmetic coding. The choice of the predictor for a particular coefficient depends on the magnitude of the horizontal and vertical gradients of neighboring coefficients. If the horizontal gradient is smaller than the vertical gradient, then prediction is performed using the left neighboring coefficient; otherwise the top neighboring coefficient is employed.

The quantized coefficients of the AC subbands are encoded using a zero-tree algorithm followed by arithmetic coding.

3.5.6   Sprite Coding

An interesting mode supported by MPEG-4 is sprite coding. A *sprite* consists of those parts of an object that are present in the scene throughout a video segment. For example, a background sprite (also referred to in the literature as a *background mosaic*) can be constructed by collecting all pels belonging to the background throughout a video segment. Note that in the case of camera panning, for example, the background sprite can be larger than the actual frames of the sequence. This still, and possibly large, image needs to be transmitted only once before transmitting the corresponding video segment. For each frame of the video segment, there is no need to encode a background VOP. Instead, a small number of parameters needs to be transmitted to allow the decoder to warp/crop the sprite and generate an appropriate background VOP. Thus, in such cases, sprite coding can achieve very high coding efficiency.

Sprite coding can operate in three modes: basic sprite coding, low-latency sprite coding, and scalable sprite coding. In *basic sprite coding* the whole sprite is encoded and transmitted to the decoder before transmitting the corresponding video segment. In *low-latency sprite coding* only part of the sprite is encoded and transmitted. This part is sufficient to be used for the first few frames of the video segment. The remaining part of the sprite is transmitted, piecewise, when required or as the bandwidth allows. In *scalable sprite coding* the sprite is encoded and transmitted progressively. In other words, a low-quality version of the sprite is encoded and transmitted first. This is then refined gradually by encoding and transmitting residuals.

3.5.7 Scalability

MPEG-4 supports both temporal and spatial scalability using multiple VOLs. For example, in the case of two VOLs, one VOL provides the base layer whereas the other VOL provides the enhancement layer.

MPEG4 uses a *generalized scalability* framework, as shown in Figure 3.15. In this framework the functionality of a block depends on the chosen type of scalability.

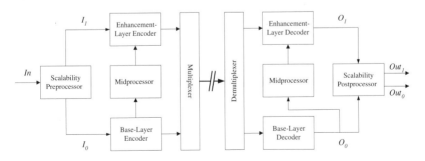

Figure 3.15: MPEG-4 generalized scalability

VOPs are input to the *scalability preprocessor*. If spatial scalability is to be performed, then this preprocessor downsamples the input VOPs to generate the base-layer VOPs forming the input to the *base-layer encoder*. The *midprocessor* takes the reconstructed base-layer VOPs and upsamples them. The difference between the original VOPs and the output of the *midprocessor* forms the enhancement-layer VOPs. Those are encoded using the *enhancement-layer encoder*. The *multiplexer* is then used to multiplex the base- and enhancement-layer bitstreams into a single bitstream. At the decoder, the *demultiplexer* is used to separate the incoming bitstream into base- and enhancement-layer bitstreams. The *scalability postprocessor* performs any necessary operations, such as upsampling the decoded base layer for display.

If, however, temporal scalability is to be performed, then the *scalability preprocessor* separates the stream of input VOPs into two substreams. One substream forms the input to the *base-layer encoder*, while the other forms the input to the *enhancement-layer encoder*. In this case, the *midprocessor* does not perform any spatial resolution conversion and simply allows the reconstructed base-layer VOPs to pass through to be used for the temporal prediction of enhancement-layer VOPs. In this case also, the *postprocessor* simply outputs the reconstructed base-layer VOPs without any conversion.

For spatial scalability, only rectangular VOPs are supported by MPEG-4. In the case of temporal scalability, however, both rectangular and arbitrary-shaped VOPs are supported. MPEG-4 provides two types of temporal scalability:

- *Type I*: The enhancement layer increases the temporal resolution of only a partial region of the base layer.

- *Type II*: The enhancement layer increases the temporal resolution of the entire region of the base layer.

3.5.8   Error Resilience

One of the main aims of MPEG-4 is to provide universal access through a wide range of environments, including error-prone environments. One of the important requirements of video communication over error-prone environments, like mobile networks, is robustness against errors. MPEG-4 provides three main tools for error resilience: resynchronization, data partitioning, and reversible VLCs.

3.5.8.1   Resynchronization

As is discussed in Chapter 9, one of the disadvantages of VLC coding is that errors in the bitstream can cause a loss of synchronization between the encoder and the decoder. One way to reduce this effect is to insert unique markers called *resynchronization codewords* in the bitstream. When an error is detected, the decoder skips the remaining bits until it finds a resynchronization codeword. This reestablishes the synchronization with the encoder, and the decoder then proceeds to decode from that point on.

Version 1 of H.263 adopts a GOB-based resynchronization approach. This means that a resynchronization codeword is inserted every time a fixed number of macroblocks (which is equal to the size of the GOB) has been encoded. Since the number of bits can vary between macroblocks, the resynchronization codewords will most likely be unevenly spaced throughout the bitstream. Therefore, certain parts of the sequence, such as high-motion areas with high bit content, will be more susceptible to errors and will also be more difficult to conceal.

MPEG-4, however, adopts a more robust approach based on video packets, as illustrated in Figure 3.16(a). In this approach each packet contains approximately the same number of bits. This means that the resynchronization codewords are almost periodic in the bitstream. Note that the header of the packet contains the necessary information (e.g., the address of the first MB in the packet and the corresponding quantization parameter) to restart decoding after reestablishing synchronization. Following the packet header is the *header extension code* (HEC). When this bit is set to "1," then additional information (e.g., timing information and VOP coding type) are included in the header.

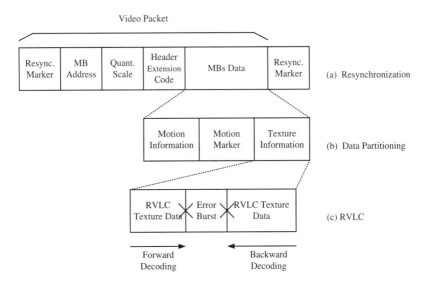

Figure 3.16: MPEG-4 error resilience tools

Such information was originally included in the VOP header. Its inclusion in the packet header as well enables the decoder to decode the packet without reference to the packet containing the VOP header. Such information can also help error detection, since it is supposed to be the same in all packets belonging to the same VOP.

Another problem with VLC coding is that errors can emulate the occurrence of start and resynchronization codewords. To reduce this effect, MPEG-4 provides a second resynchronization approach called *fixed-interval synchronization*. In this approach, VOP start codes and packet resynchronization codewords appear only at legal fixed-interval locations in the bitstream. Thus, only codewords at those legal locations will be used by the decoder to reestablish synchronization.

3.5.8.2 Data Partitioning

In some cases, an error occurs well before the point in the bitstream at which the error is detected. Therefore, when an error is detected, all bits between the resynchronization codeword prior to the error detection point and the resynchronization codeword where synchronization is reestablished are typically discarded. If the decoder can localize the error more effectively, then the performance of error concealment techniques (discussed in Chapter 9) can be improved.

MPEG-4 uses data partitioning to further improve the ability of the decoder to localize errors. In this approach the bitstream between two resynchronization codewords is divided into smaller logical units. Each logical unit contains one type of information for *all* MBs belonging to the same packet. For example, in Figure 3.16(b) the motion information for all the MBs in the packet is encoded first, followed by a motion marker and then the texture information for all the MBs in the packet. In the non-data-partitioned case, if an error occurs in the texture information, then the header, motion, and texture information will all be discarded. In the data-partitioned case, however, if an error occurs in the texture information, then only the texture information will be discarded, and the motion marker will be used to locate and recover the header and motion information. Temporal concealment (described in Chapters 9 and 10) can then use this recovered information to conceal the corrupted MBs from the reference VOP.

3.5.8.3 Reversible VLCs

As already discussed, when an error is detected in the bitstream, the bits between the surrounding resynchronization codewords are discarded, and the decoder skips to the next resynchronization codeword and proceeds decoding from there. In MPEG-4, however, texture information is encoded using

RVLCs, as illustrated in Figure 3.16(c). In this case, when the decoder jumps to the next resynchronization codeword, instead of discarding all preceding bits, the decoder can start decoding in the reverse direction to recover and utilize some of those bits.

3.5.9 Profiles and Levels

As already discussed, profiles and levels provide a means of defining subsets of the syntax and semantics of a standard. This in turn provides a means of defining the decoder capabilities required to decode a particular bitstream. Profiles and levels are used to define conformance points that facilitate bitstream interchange among different applications.

In MPEG-4, *object types* are used to define profiles. An object type defines a subset of MPEG-4 tools that provides a single or a group of functionalities. There are six natural video object types: simple, core, main, simple scalable, N-bit, and still scalable texture. For example, the *main object type* includes the following subset of tools: basic (I- and P-VOP, coefficient prediction, 4-MV, and unrestricted MV), error resilience, short header, B-VOP, Methods 1 and 2 for quantization, P-VOP-based temporal scalability, binary shape, gray shape, interlace, and sprite.

A *profile* is a defined subset of the entire bitstream syntax. MPEG-4 defines six natural video profiles: simple, core, main, simple scalable, N-bit, and scalable texture. Each profile is defined in terms of video object types. For example, the *main profile* includes the following object types: simple, core, main, and scalable still texture.

A *level* within a profile is a defined set of constraints imposed on parameters in the bitstream that relate to the tools of that profile. For example, *level 1* (L1) of the *simple profile* has a typical session size of QCIF, a maximum total number of objects of 4, and a maximum bitrate of 64 kbits/s.

Part II
Coding Efficiency

The radio spectrum is a limited and scarce resource. This puts very stringent limits on the bandwidth available for a mobile channel. Given the enormous amount of data generated by video, the use of efficient coding techniques is vital.

One of the most important factors that decide the coding efficiency of a video codec is the motion estimation and compensation technique. This part contains three chapters. Chapter 4 covers some basic motion estimation methods. It starts by introducing some of the fundamentals of motion estimation. It then reviews some basic motion estimation methods, with particular emphasis on the widely used block-matching methods. The chapter then presents the results of a comparative study between the different methods. The chapter also investigates the efficiency of block-matching motion estimation at very low bit rates, typical of mobile video communication. The aim is to decide if the added complexity of this process is justifiable, in terms of an improved coding efficiency, at such bit rates.

Chapter 5 investigates the performance of the more advanced warping-based motion estimation methods. The chapter starts by describing a general warping-based motion estimation method. It then considers some important parameters, like the shape of the patches, the spatial transformation used, and the node-tracking algorithm. The chapter then assesses the suitability of warping-based methods for mobile video communications. In particular, the chapter compares the coding efficiency and the computational complexity of such methods to those of block-matching methods.

Chapter 6 investigates the performance of another advanced motion estimation method, called multiple-reference motion-compensated prediction (MR-MCP). The chapter starts by briefly reviewing multiple-reference motion estimation methods. It then concentrates on the long-term memory motion-compensated prediction (LTM-MCP) technique. The chapter investigates the prediction gains and the coding efficiency of this technique at very low bit rates. The primary aim is to decide if the added complexity, increased motion overhead, and increased memory requirements of this technique are justifiable at such bit rates. The chapter also investigates the properties of multiple-reference block motion fields and compares them to those of single-reference fields.

Chapter 4

Basic Motion Estimation Techniques

4.1 Overview

Motion estimation is an important process in a wide range of disciplines and applications, such as image sequence analysis, computer vision, target tracking, and video coding. Different disciplines and applications have different requirements and may, therefore, use different motion estimation techniques.

This chapter reviews some basic motion estimation techniques developed specifically for video coding. It then carries out a comparative study between the different techniques. The chapter also presents the results of an investigation into the efficiency of block-matching motion estimation at very low bit rates. In particular, the investigation shows that the added complexity of this process is justifiable at such bit rates.

Section 4.2 gives a brief introduction to the basics of motion estimation. Sections 4.3–4.6 briefly review the differential, pel-recursive, frequency-domain, and block-matching motion estimation methods. Section 4.7 presents the results of a comparative study of the reviewed techniques, whereas Section 4.8 investigates the efficiency of motion estimation at very low bit rates. The chapter concludes with a discussion in Section 4.9.

4.2 Motion Estimation

As already discussed in Chapter 2 (Section 2.7.2), the most commonly used video coding method is motion-compensated coding. In the first stage of this method, called *motion estimation* (ME), the motion of objects between a reference frame and the current frame is estimated. This motion information is then used in the second stage, called *motion compensation* (MC), to move the objects of the reference frame to provide a prediction for the current frame. The prediction error, called the *displaced-frame difference* (DFD), is encoded instead of the current frame itself. The estimated motion information also has to be transmitted, unless the decoder can estimate it from previously decoded information. This section introduces the basics of motion estimation. It defines and formulates the motion estimation problem and describes the main approaches and models used to solve this problem. Examples of such solutions will be discussed in subsequent sections.

4.2.1 Projected Motion and Apparent Motion

In video, the 3-D motion of objects in space is projected as 2-D motion onto the image plane. This 2-D motion, called *projected motion*, is illustrated in Figure 4.1. Thus, motion estimation may refer to the process of estimating image-plane 2-D motion or object-space 3-D motion. Note that the two are not equivalent. In fact, 2-D motion estimation is usually the first step toward 3-D motion estimation. This chapter considers *2-D motion estimation* only. For 3-D motion estimation, the reader is referred to Ref. 10.

In video coding, motion is estimated by observing the spatiotemporal variation of intensity between frames. This is called the *apparent motion*. In the ideal case, apparent motion is equivalent to true projected motion. In practice, however, this is not always the case. For example, when a circle with uniform

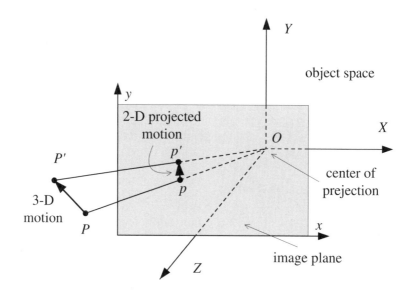

Figure 4.1: Projected motion

intensity rotates about its center, it has a rotational projected motion but zero apparent motion. Another example is a still object with change of illumination between frames. Although the object has zero projected motion, the change in illumination will result in some apparent motion. Hereafter, unless otherwise stated, the term *motion* will be used to refer to apparent motion rather than true projected motion.

Two-dimensional motion can be represented in terms of either 2-D displacement vectors, $\mathbf{d} = [d_x, d_y]^T$, or 2-D instantaneous velocity vectors, $\mathbf{v} = [v_x, v_y]^T = [\frac{dx}{dt}, \frac{dy}{dt}]^T$. A set of such vectors representing motion in a frame is called the *motion field* of the frame. The two representations are called the *displacement field* and the *velocity field* in the case of projected motion, or the *correspondence field* and the *optical flow field* in the case of apparent motion. However, in the video coding literature, it has become a convention to ignore this distinction and to use the terms *displacement field* and *velocity field* to refer to the apparent correspondence field and optical flow field, respectively. Hereafter, this convention will be adopted. Furthermore, this book uses the displacement field representation rather than the velocity field representation. Thus, the term *motion field* will always refer to the apparent correspondence field and the term *motion vector* will always refer to a displacement vector within this field.

4.2.2 Problem Formulation

Two-dimensional apparent motion can be attributed to three main causes. The first cause is global, or camera, motion. Even when there is no object motion in the frame, the motion of the camera induces a global motion. The second cause is local motion. This is the intrinsic motion of the objects in the scene. The third cause is illumination changes. Even when there is no object motion in the scene, changes in lighting conditions influence apparent motion.

All techniques considered in this chapter make no distinction between global and local motions, and they do not take into account illumination changes. Thus, they assume that global motion is taken into account through local motion and that the impact of illumination changes can be ignored. It should be pointed out, however, that some other techniques use a two-stage global/local motion estimation, e.g., Ref. 77, or estimate illumination changes, e.g., Ref. 78.

The 2-D apparent motion estimation problem can be formulated as a forward or a backward estimation problem depending on the temporal location of the reference frame with respect to the current frame.

In *backward motion estimation*, a pel $\mathbf{s} = [x, y]^T$ in the current frame at time t is related to a pel in a previous reference frame at time $t - \Delta t$ by

$$f_t(\mathbf{s}) = f_{t-\Delta t}(\mathbf{s} - \mathbf{d}(\mathbf{s})). \tag{4.1}$$

In *forward motion estimation*, however, the same pel is related to a pel in a future reference frame at time $t + \Delta t$ by

$$f_t(\mathbf{s}) = f_{t+\Delta t}(\mathbf{s} + \mathbf{d}(\mathbf{s})). \tag{4.2}$$

The aim of motion estimation is to find the motion vector $\mathbf{d}(\mathbf{s}) = [d_x(\mathbf{s}), d_y(\mathbf{s})]^T$. Note that $\mathbf{d}(\mathbf{s})$ is not necessarily a full-pel accurate motion vector. Thus, a motion estimation technique may need to access intensity values at nonsampling locations in the reference frame. This is achieved using interpolation techniques like nearest-neighbor, bilinear, and cubic interpolation. In this book, bilinear interpolation is employed because of its good compromise between interpolation quality and computational complexity. It is defined as

$$f(x, y) = (1 - x_f)(1 - y_f)f(x_i, y_i) + x_f(1 - y_f)f(x_i + 1, y_i)$$

$$+ (1 - x_f)y_f f(x_i, y_i + 1) + x_f y_f f(x_i + 1, y_i + 1), \tag{4.3}$$

where (x_i, y_i) and (x_f, y_f) are, respectively, the integer and fractional parts of the pel coordinates (x, y).

Care should be taken when interpreting the terms *forward* and *backward*. The two terms can be used to refer to either the motion estimation process or the motion compensation process. A forward motion estimation process corresponds to a backward motion compensation process, and vice versa. Note that forward motion estimation is associated with a coding delay. Thus, most video coding standards employ backward estimation (i.e., forward compensation), although forward estimation is sometimes employed (e.g., in B-frames in MPEG1–2 and PB-frames in H.263).

4.2.3   An Ill-Posed Problem

The preceding formulation of the motion estimation problem indicates that it is an ill-posed problem.[1] It suffers from the following problems [10]:

- Existence of solution: For example, no motion can be estimated for covered/uncovered background pels. This is known as the *occlusion problem*.

- Uniqueness of solution: At each pel, $\mathbf{s}$, the number of unknown independent variables (d_x and d_y) is twice the number of equations, (4.1) or (4.2). This is known as the *aperture problem*.

---

[1]A problem is called ill-posed if a unique solution does not exist and/or the solution does not continuously depend on the data [79].

- Continuity of solution: The motion estimate is highly sensitive to the presence of noise.

Because of this ill-posed nature of the problem, motion estimation algorithms use additional assumptions about the structure of the motion field. Such assumptions are referred to as *motion models*. They can be deterministic or probabilistic, parametric or nonparametric, as will be discussed in the following subsections.

4.2.4 Deterministic and Probabilistic Models

In a deterministic model, motion is seen as an unknown deterministic quantity. By maximizing the probability of the observed video sequence with respect to the unknown motion, this deterministic quantity can be estimated. The corresponding estimator is usually referred to as a *maximum likelihood* (ML) estimator. All motion estimation methods discussed in this chapter follow this deterministic approach.

In a probabilistic (or Bayesian) model, motion is seen as a random variable. Thus, the ensemble of motion vectors forms a random field. This field is usually modeled using a *Markov random field* (MRF). Given this model, motion estimation can be formulated as a *maximum a posteriori probability* (MAP) estimation problem. This problem can be solved using optimization techniques like simulated annealing, iterated conditional modes, mean field annealing, and highest confidence first. For a detailed description of Bayesian motion estimation methods, the reader is referred to Ref. 10.

4.2.5 Parametric and Nonparametric Models

In a parametric model, motion is represented by a set of *motion parameters*. Thus, the problem of motion estimation becomes a problem of estimating the motion parameters rather than the motion field itself. Since 2-D motion results from the projection of 3-D motion onto the image plane, a parametric 2-D motion model is usually derived from models describing 3-D motion, 3-D surfaces, and the projection geometry. For example, the assumptions of a planar 3-D surface moving in space according to a 3-D affine model and projected onto the image plane using an orthographic projection[2] results in a 2-D 6-parameter affine model. Different assumptions lead to different 2-D models. The 2-D models can be as complex as a quadratic 12-parameter model

---

[2]In an orthographic projection, it is assumed that all rays from a projected 3-D object to the image plane travel parallel to each other [10].

or as simple as a translational 2-parameter model (which is used in block-matching) [80]. Note that with parametric models, the constraint to regularize the ill-posed motion estimation problem is implicitly included in the motion model.

In nonparametric models, however, an explicit constraint (e.g., the smoothness of the motion field) is introduced to regularize the ill-posed problem of motion estimation.

4.2.6   Region of Support

An important parameter in motion estimation is the region of support. This is the set of pels to which the motion model applies. A region of support can be as large as a frame or as small as a single pel, it can be of fixed size or of variable size, and it can have a regular shape or an arbitrary shape.

Large regions of support result in a small motion overhead but may suffer from the *accuracy problem*. This means that pels within the region belong to different objects moving in different directions. Thus, the estimated motion parameters will not be accurate for some or all of the pels within the region.

The accuracy problem can be overcome by using small regions of support. This is, however, at the expense of an increase in motion overhead. Small support regions may also suffer from the *ambiguity problem*. This means that several patterns similar to the region may appear at multiple locations within the reference frame. This may lead to incorrect motion parameters.

4.3   Differential Methods

Differential methods are among the early approaches for estimating the motion of objects in video sequences. They are based on the relationship between the spatial and the temporal changes of intensity.

Differential methods were first proposed by Limb and Murphy in 1975 [81]. In their method, they use the magnitude of the temporal frame difference, FD, over a moving area, $\mathcal{A}$, to measure the speed of this area. To remove dependence on the area size, this measure is normalized by the horizontal, HD, or vertical, VD, spatial pel differences. Thus the estimated motion vector is given by

$$\hat{\mathbf{d}} = \begin{bmatrix} \hat{d}_x \\ \hat{d}_y \end{bmatrix} = \begin{bmatrix} \dfrac{\sum_{\mathbf{s}\in\mathcal{A}}\mathrm{FD}(\mathbf{s})\mathrm{sign}(\mathrm{HD}(\mathbf{s}))}{\sum_{\mathbf{s}\in\mathcal{A}}|\mathrm{HD}(\mathbf{s})|} \\ \dfrac{\sum_{\mathbf{s}\in\mathcal{A}}\mathrm{FD}(\mathbf{s})\mathrm{sign}(\mathrm{VD}(\mathbf{s}))}{\sum_{\mathbf{s}\in\mathcal{A}}|\mathrm{VD}(\mathbf{s})|} \end{bmatrix}, \tag{4.4}$$

where

$$\text{sign}(z) = \begin{cases} \frac{z}{|z|}, & \text{if } |z| \geq \text{threshold}, \\ 0, & \text{otherwise}, \end{cases} \tag{4.5}$$

$$\text{FD}(\mathbf{s}) = f_t(\mathbf{s}) - f_{t-\Delta t}(\mathbf{s}), \tag{4.6}$$

$$\text{HD}(\mathbf{s}) = \frac{1}{2}[f_t(x+1, y) - f_t(x-1, y)], \tag{4.7}$$

and

$$\text{VD}(\mathbf{s}) = \frac{1}{2}[f_t(x, y+1) - f_t(x, y-1)]. \tag{4.8}$$

The theoretical basis of differential methods were established later by Cafforio and Rocca in 1976 [82]. They start with the basic definition of the frame difference, Equation (4.6), and they rewrite it as

$$\text{FD}(\mathbf{s}) = f_t(\mathbf{s}) - f_{t-\Delta t}(\mathbf{s})$$

$$= f_t(\mathbf{s}) - f_t(\mathbf{s} + \mathbf{d}). \tag{4.9}$$

For small values of $\mathbf{d}$, the right-hand side of Equation (4.9) can be replaced by its Taylor series expansion about $\mathbf{s}$, as follows:

$$\text{FD}(\mathbf{s}) = -\mathbf{d}^T \nabla_\mathbf{s} f_t(\mathbf{s}) + \text{higher-order terms}, \tag{4.10}$$

where $\nabla_\mathbf{s} = [\frac{\partial}{\partial x}, \frac{\partial}{\partial y}]^T$ is the spatial gradient with respect to $\mathbf{s}$. Ignoring the higher-order terms and assuming that motion is constant over an area $\mathscr{A}$, linear regression can be used to obtain the minimum mean square estimate of $\mathbf{d}$ as

$$\hat{\mathbf{d}} = -\left[\sum_{\mathbf{s} \in \mathscr{A}} \nabla_\mathbf{s} f_t(\mathbf{s}) \nabla_\mathbf{s}^T f_t(\mathbf{s}) \right]^{-1} \left[\sum_{\mathbf{s} \in \mathscr{A}} \text{FD}(\mathbf{s}) \nabla_\mathbf{s} f_t(\mathbf{s}) \right]. \tag{4.11}$$

Note that this equation is highly dependent on the spatial gradient, $\nabla_\mathbf{s}$. For this reason, differential methods are also known as *gradient methods*. Using the approximation $\nabla_\mathbf{s} f_t(\mathbf{s}) \approx [\text{HD}(\mathbf{s}), \text{VD}(\mathbf{s})]^T$, Equation (4.11) reduces to

$$\hat{\mathbf{d}} = -\left[\begin{array}{cc} \sum_{\mathbf{s} \in \mathscr{A}} \text{HD}^2(\mathbf{s}) & \sum_{\mathbf{s} \in \mathscr{A}} \text{HD}(\mathbf{s}) \cdot \text{VD}(\mathbf{s}) \\ \sum_{\mathbf{s} \in \mathscr{A}} \text{HD}(\mathbf{s}) \cdot \text{VD}(\mathbf{s}) & \sum_{\mathbf{s} \in \mathscr{A}} \text{VD}^2(\mathbf{s}) \end{array} \right]^{-1}$$

$$\times \left[\begin{array}{c} \sum_{\mathbf{s} \in \mathscr{A}} \text{FD}(\mathbf{s}) \cdot \text{HD}(\mathbf{s}) \\ \sum_{\mathbf{s} \in \mathscr{A}} \text{FD}(\mathbf{s}) \cdot \text{VD}(\mathbf{s}) \end{array} \right]. \tag{4.12}$$

By ignoring the cross terms (i.e., $\sum_{s \in \mathcal{A}} HD(s) \cdot VD(s) \approx 0$), it can be shown that the general analytical solution of Cafforio and Rocca (Equation (4.12)) reduces to the simple heuristic solution of Limb and Murphy (Equation (4.4)).

The main assumption in deriving the differential estimate of Equation (4.12) using Taylor series expansion is that the motion vector **d** is small. As **d** increases, the quality of the approximation becomes poor. Thus, the main drawback of differential methods is that they can only be used to measure small motion displacements (up to about ± 3 pels). A number of methods have been proposed to overcome this problem, like, for example, the iterative method of Yamaguchi [83]. In this method, an initial motion vector is first estimated, using Equation (4.12), between a block in the current frame and a corresponding block in the same location in the reference frame. In the next iteration, the position of the matched block in the reference frame is shifted by the initial motion vector, and then the differential method is applied again to produce a second estimate. This second estimate acts as a correction term for the initial estimate. This process of shift and estimation continues until the correction term becomes adequately small.

Another drawback of differential methods is that the spatial gradient operator, ∇_s, is sensitive to data noise. This can be reduced by using a larger set of data in its calculation.

There are also cases where differential methods can fail [84]. For example, in smooth areas the gradient is approximately equal to zero and the matrix in Equation (4.12) becomes singular. Also, when motion is parallel to edges in the image, i.e., $\mathbf{d}^T \nabla_s \approx 0$, the frame difference, Equation (4.10), becomes zero, giving a wrong displacement of zero. Such problems may be partially solved by increasing the data area, but this may give rise to the accuracy problem.

4.4 Pel-Recursive Methods

Given a function $g(\mathbf{r})$ of several unknowns $\mathbf{r} = [r_1, \dots, r_n]^T$, the most straightforward way to minimize it is to calculate its partial derivatives with respect to each unknown, set them equal to 0, and solve the resulting simultaneous equations. This is called *gradient-based optimization* and can be represented in vector form as

$$\nabla_{\mathbf{r}} g(\mathbf{r}) = 0. \tag{4.13}$$

In cases where the function $g(\mathbf{r})$ cannot be represented in closed form and/or the set of simultaneous Equations (4.13) cannot be solved, numerical iterative methods are employed.

One of the simplest numerical methods is the steepest-descent method. Since the gradient vector points in the direction of the maximum, this method updates the present estimate, $\hat{\mathbf{r}}^i$, of the location of the minimum in the direction of the negative gradient, to obtain a new improved estimate

$$\hat{\mathbf{r}}^{i+1} = \hat{\mathbf{r}}^i - \alpha \nabla_{\mathbf{r}} g(\hat{\mathbf{r}}^i), \tag{4.14}$$

where $\alpha > 0$ is an update step size and i is the iteration index.

Pel-recursive methods are based on an iterative gradient-based minimization of the prediction error. They were first proposed by Netravali and Robbins in 1979 [85]. In their algorithm, they use a steepest-descent approach to iteratively minimize the square of the displaced-frame difference, DFD($\mathbf{s}, \mathbf{d}$), with respect to the displacement vector, $\mathbf{d}$. Thus

$$g(\mathbf{r}) = \mathrm{DFD}^2(\mathbf{s}, \mathbf{d}), \tag{4.15}$$

where

$$\mathrm{DFD}(\mathbf{s}, \mathbf{d}) = f_t(\mathbf{s}) - f_{t-\Delta t}(\mathbf{s} - \mathbf{d}). \tag{4.16}$$

Substituting Equation (4.15) into Equation (4.14) and setting $\alpha = \frac{\varepsilon}{2}$ gives

$$\hat{\mathbf{d}}^{i+1} = \hat{\mathbf{d}}^i - \frac{\varepsilon}{2} \nabla_{\mathbf{d}} \mathrm{DFD}^2(\mathbf{s}, \hat{\mathbf{d}}^i). \tag{4.17}$$

Now,

$$\nabla_{\mathbf{d}} \mathrm{DFD}^2(\mathbf{s}, \mathbf{d}) = 2\,\mathrm{DFD}(\mathbf{s}, \mathbf{d})\,\nabla_{\mathbf{d}} \mathrm{DFD}(\mathbf{s}, \mathbf{d})$$

$$= 2\,\mathrm{DFD}(\mathbf{s}, \mathbf{d})\,\nabla_{\mathbf{d}}[f_t(\mathbf{s}) - f_{t-\Delta t}(\mathbf{s} - \mathbf{d})]$$

$$= 2\,\mathrm{DFD}(\mathbf{s}, \mathbf{d})\,\nabla_{\mathbf{s}} f_{t-\Delta t}(\mathbf{s} - \mathbf{d}). \tag{4.18}$$

Substituting Equation (4.18) into Equation (4.17) gives

$$\hat{\mathbf{d}}^{i+1} = \hat{\mathbf{d}}^i - \varepsilon \mathrm{DFD}(\mathbf{s}, \hat{\mathbf{d}}^i)\nabla_{\mathbf{s}} f_{t-\Delta t}(\mathbf{s} - \hat{\mathbf{d}}^i), \tag{4.19}$$

where the spatial gradient $\nabla_{\mathbf{s}} f_{t-\Delta t}(\mathbf{s} - \hat{\mathbf{d}}^i)$ can be approximated by Equations (4.7) and (4.8) but evaluated at a displaced location $(\mathbf{s} - \mathrm{NINT}[\hat{\mathbf{d}}^i])$ in the reference frame. As in differential methods, this estimate is highly dependent on the spatial gradient. For this reason, pel-recursive methods are sometimes considered a subset of gradient or differential methods.

The iterative approach of Equation (4.19) is normally applied on a pel-by-pel basis, leading to a *dense* motion field, $\hat{\mathbf{d}}(\mathbf{s})$. Iterations may proceed along a scanning line, from line to line, or from frame to frame. In order to smooth out the effect of noise, the update term can be evaluated over an area $\mathscr{A} = \{\mathbf{s}_1, \ldots, \mathbf{s}_p\}$ as follows:

$$\hat{\mathbf{d}}^{i+1} = \hat{\mathbf{d}}^i - \varepsilon \sum_{j=1}^{p} W_j\,\mathrm{DFD}(\mathbf{s}_j, \hat{\mathbf{d}}^i)\nabla_{\mathbf{s}} f_{t-\Delta t}(\mathbf{s}_j - \hat{\mathbf{d}}^i), \tag{4.20}$$

where $W_j \geq 0$ and $\sum_{j=1}^{p} W_j = 1$. Netravali and Robbins also proposed a simplified expression for hardware implementation:

$$\hat{\mathbf{d}}^{i+1} = \hat{\mathbf{d}}^i - \varepsilon \, \text{sign}[\text{DFD}(\mathbf{s}, \hat{\mathbf{d}}^i)] \, \text{sign}[\nabla_{\mathbf{s}} f_{t-\Delta t}(\mathbf{s} - \hat{\mathbf{d}}^i)]. \tag{4.21}$$

The convergence of this method is highly dependent on the constant step size ε. A high value of ε leads to quick convergence but less accuracy, whereas a small value of ε leads to slower convergence but more accurate estimates. Thus, a compromise between the two is desired. A number of algorithms have been reported to improve the performance of pel-recursive algorithms, e.g., Ref. 86. Most of them are based on the idea of substituting the constant step size ε by a variable step size to achieve better adaptation to the local image statistics and, consequently, faster convergence and higher accuracy. A good review of such methods with comparative results can be found in Ref. 87.

The dense motion field of pel-recursive methods can overcome the accuracy problem. This is, however, at the expense of a large motion overhead. To overcome this drawback, the update term from one iteration to the other can be based on previously transmitted data only. In this case, the decoder can estimate the same displacements generated at the encoder, and no motion information needs to be transmitted. A disadvantage of this causal approach, however, is that it constrains the method and reduces its prediction capability. In addition, it increases the complexity of the decoder.

Another disadvantage of pel-recursive methods is that they can easily converge to local minima within the error surface. In addition, smooth intensity regions, discontinuities within the motion field, and large displacements cannot be efficiently handled [55].

4.5 Frequency-Domain Methods

Frequency-domain motion estimation methods are based on the *Fourier transform* (FT) property that a translational displacement in the spatial domain corresponds to a linear phase shift in the frequency domain. Thus, assuming that the image intensities of the current frame, f_t, and the reference frame, $f_{t-\Delta t}$, differ over a moving area, $\mathscr{A}$, only due to a translational displacement, (d_x, d_y), then

$$f_t(x, y) = f_{t-\Delta t}(x - d_x, y - d_y), \quad (x, y) \in \mathscr{A}. \tag{4.22}$$

Taking the FT of both sides with respect to the spatial variables (x, y) gives the following frequency-domain equation in the frequency variables (w_x, w_y):

$$F_t(w_x, w_y) = F_{t-\Delta t}(w_x, w_y) e^{j(-w_x d_x - w_y d_y)}, \tag{4.23}$$

where F_t and $F_{t-\Delta t}$ are the FTs of the current and reference frames, respectively. In Ref. 88, Haskell noticed this relationship but did not propose an algorithm to recover the displacement from the phase shift.

If we define $\Delta\phi(w_x, w_y)$ as the phase difference between the FT of the current frame and that of the reference frame, then

$$
\begin{aligned}
e^{j\Delta\phi(w_x, w_y)} &= e^{j[\phi_t(w_x, w_y) - \phi_{t-\Delta t}(w_x, w_y)]} \\
&= e^{j\phi_t(w_x, w_y)} \cdot e^{-j\phi_{t-\Delta t}(w_x, w_y)} \\
&= \frac{F_t(w_x, w_y)}{|F_t(w_x, w_y)|} \cdot \frac{F^*_{t-\Delta t}(w_x, w_y)}{|F^*_{t-\Delta t}(w_x, w_y)|},
\end{aligned}
\tag{4.24}
$$

where ϕ_t and $\phi_{t-\Delta t}$ are the phase components of F_t and $F_{t-\Delta t}$, respectively, and the superscript $*$ indicates the complex conjugate. If we define $c_{t, t-\Delta t}(x, y)$ as the inverse FT of $e^{j\Delta\phi(w_x, w_y)}$, then

$$
\begin{aligned}
c_{t, t-\Delta t}(x, y) &= \mathscr{F}^{-1}\{e^{j\Delta\phi(w_x, w_y)}\} \\
&= \mathscr{F}^{-1}\{e^{j\phi_t(w_x, w_y)} \cdot e^{-j\phi_{t-\Delta t}(w_x, w_y)}\} \\
&= \mathscr{F}^{-1}\{e^{j\phi_t(w_x, w_y)}\} \otimes \mathscr{F}^{-1}\{e^{-j\phi_{t-\Delta t}(w_x, w_y)}\},
\end{aligned}
\tag{4.25}
$$

where $\otimes$ is the 2-D convolution operation. In other words, $c_{t, t-\Delta t}(x, y)$ is the cross-correlation of the inverse FTs of the phase components of F_t and $F_{t-\Delta t}$. For this reason, $c_{t, t-\Delta t}(x, y)$ is known as the *phase correlation function*. The importance of this function becomes apparent if it is rewritten in terms of the phase difference in Equation (4.23):

$$
\begin{aligned}
c_{t, t-\Delta t}(x, y) &= \mathscr{F}^{-1}\{e^{j\Delta\phi(w_x, w_y)}\} \\
&= \mathscr{F}^{-1}\{e^{j(-w_x d_x - w_y d_y)}\} \\
&= \delta(x - d_x, y - d_y).
\end{aligned}
\tag{4.26}
$$

Thus, the phase correlation surface has a distinctive impulse at (d_x, d_y). This observation is the basic idea behind the *phase correlation* motion estimation method. In this method, Equation (4.24) is used to calculate $e^{j\Delta\phi(w_x, w_y)}$, the inverse FT is then applied to obtain $c_{t, t-\Delta t}(x, y)$, and the location of the impulse in this function is detected to estimate (d_x, d_y).

In practice, the impulse in the phase correlation function degenerates into one or more peaks. This is due to many factors, like the use, in digital images, of the discrete Fourier transform (DFT) instead of the FT, the presence of more than one moving object within the considered area $\mathscr{A}$, and the presence of noise. In particular, the use of the 2-D DFT instead of the 2-D FT results in the following effects [10]:

- *The boundary effect*: In order to obtain a perfect impulse, the translational displacement must be cyclic. In other words, objects disappearing at one end of the moving area must reappear at the other end. In practice this does not happen, which leads to the degeneration of the impulse into peaks. Furthermore, the DFT assumes periodicity in both directions. In practice, however, discontinuities occur from left to right and from top to bottom, introducing spurious peaks.

- *Spectral leakage*: In order to obtain a perfect impulse, the translational displacement must correspond to an integer multiple of the fundamental frequency. In practice, noninteger motion vectors may not satisfy this condition, leading to the well-known spectral leakage phenomenon [89], which degenerates the impulse into peaks.

- *Displacement wrapping*: The 2-D DFT is periodic with the area size (N_x, N_y). Negative estimates will be wrapped and will appear as positive displacements. To accommodate negative displacements, the estimated displacement needs to be unwrapped as follows [10]:

$$d_i = \begin{cases} \hat{d}_i & \text{if } |\hat{d}_i| \leq \frac{N_i}{2} \text{ and } N_i \text{ is even} \\ & \text{or if } |\hat{d}_i| \leq \frac{N_i-1}{2} \text{ and } N_i \text{ is odd,} \\ \hat{d}_i - N_i, & \text{otherwise.} \end{cases} \qquad (4.27)$$

This means that the range of estimates is limited to $[\frac{-N_i}{2} + 1, \frac{N_i}{2}]$ for N_i even.

The phase correlation motion estimation method was first reported by Kuglin and Hines in 1975 [90]. It was later extensively studied by Thomas [91]. In his study, Thomas analyzed the properties of the phase correlation function. He suggested using a weighting function to smooth the correlation surface and suppress spurious peaks. He also proposed a second stage to the method, in which smaller moving areas are used and more than one dominant peak from the first stage are considered and compared. Girod [92] augmented this by a third stage, in which the estimated integer-pel motion displacement is refined to subpel accuracy.

The phase correlation method has a number of desirable properties. It has a small computational complexity, especially with the use of *fast Fourier transforms* (FFTs). In addition, it is relatively insensitive to illumination changes because shifts in the mean value or multiplication by a constant do not affect the Fourier phase. Furthermore, the method can detect multiple moving objects, because they appear as multiple peaks in the correlation surface. In addition to its use in video coding, the phase correlation method has been successfully incorporated into commercial standards conversion equipment [93].

There are few other frequency-domain motion estimation methods. For example, Chou and Hang [94] analyzed frequency-domain motion estimation in both noise-free and noisy situations. Their analysis is very similar to the noise analysis in phase or frequency modulation systems, and it provides insights into the performance limits of motion estimation. They formulated frequency-domain motion estimation as a set of simultaneous equations, which they solved using a modified *least-mean-square* (LMS) algorithm. The resulting algorithm is known as the *frequency component method*. It provides more reliable estimates than the phase correlation method, particularly for noisy sequences. Young and Kingsbury [95] proposed a frequency-domain method based on the complex lapped transform. Koc and Liu [96] used the pseudophase hidden in the DCT transform to propose a DCT-based frequency-domain motion estimation method. The algorithm has a low computational complexity and was later extended to achieve interpolation-free subpel accuracy [97].

4.6 Block-Matching Methods

Block-matching motion estimation (BMME) is the most widely used motion estimation method for video coding. Interest in this method was initiated by Jain and Jain in 1981 [54]. In their *block-matching algorithm* (BMA), the current frame, f_t, is first divided into blocks of $M \times N$ pels. The algorithm then assumes that all pels within the block undergo the same translational movement. Thus, the same motion vector, $\mathbf{d} = [d_x, d_y]^T$, is assigned to all pels within the block. This motion vector is estimated by searching for the *best-match* block in a larger *search window* of $(M + 2d_{m_x}) \times (N + 2d_{m_y})$ pels centered at the same location in a reference frame, $f_{t-\Delta t}$, where d_{m_x} and d_{m_y} are the maximum allowed motion displacements in the horizontal and vertical directions, respectively. This process is illustrated in Figure 4.2 and can be formulated as follows:

$$(\hat{d}_x, \hat{d}_y) = \arg \min_{i,j} \mathrm{BDM}(i,j), \text{ where } |i| \leq d_{m_x} \text{ and } |j| \leq d_{m_y}, \quad (4.28)$$

and $\mathrm{BDM}(i,j)$ is a *block distortion measure* that measures the quality of match between the block in the current frame and a corresponding candidate block in the reference frame shifted by a displacement (i,j). It is very common to use square blocks of $N \times N$ pels and a maximum motion displacement of $\pm d_m$ in both directions. When Equation (4.28) is evaluated for all possible (i,j)

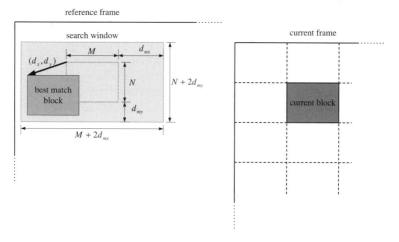

Figure 4.2: Block-matching motion estimation

displacements (i.e., for all possible candidate blocks in the search window), the BMA is referred to as the *full-search* (FS) algorithm.

Since its introduction, BMME has attracted considerable attention, and many refinements to the basic BMA have been proposed. In the following subsections, different parameters of the BMA are introduced and their impact on performance is evaluated. A number of refinements to the basic BMA are also examined.

4.6.1   Matching Function

The matching function (or the BDM) can be any function that measures the distortion or the match between the block, $\mathcal{B}$, in the current frame and the displaced candidate block in the reference frame. The choice of a suitable BDM is very important, for it impacts both the prediction quality and the computational complexity of the algorithm.

One possible matching function is the *normalized cross-correlation function*[3] (NCCF), defined as

$$\text{NCCF}(i,j) = \frac{\sum_{(x,y)\in\mathcal{B}} f_t(x,y) \cdot f_{t-\Delta t}(x-i,y-j)}{\sqrt{\sum_{(x,y)\in\mathcal{B}} f_t^2(x,y)} \cdot \sqrt{\sum_{(x,y)\in\mathcal{B}} f_{t-\Delta t}^2(x-i,y-j)}}. \qquad (4.29)$$

---

[3]The NCCF is a measure of the correlation between two blocks rather than the distortion between them. Thus, when used in BMA, the minimization process in Equation (4.28) becomes a maximization process.

Since the motion estimation process aims at minimizing the DFD signal, a natural choice for the matching function is the *mean squared error*, which is often formulated as the *sum of squared differences* (SSD):

$$\text{SSD}(i,j) = \sum_{(x,y)\in\mathscr{B}} (f_t(x, y) - f_{t-\Delta t}(x - i, y - j))^2. \tag{4.30}$$

A very similar matching function is the *sum of absolute differences* (SAD):

$$\text{SAD}(i,j) = \sum_{(x,y)\in\mathscr{B}} |f_t(x, y) - f_{t-\Delta t}(x - i, y - j)|. \tag{4.31}$$

To compare the performance of these matching functions, a full-pel full-search BMA was implemented. The algorithm uses 16×16 blocks and a maximum allowed motion displacement of ± 15 pels in both directions. In this algorithm, motion is estimated and compensated using original previous frames, and motion vectors are restricted so that they do not point outside the reference frame. Motion vectors are encoded using the median predictor and the VLC table of the H.263 standard. Unless otherwise stated, all subsequent results in this chapter use the same simulation conditions. Figure 4.3 compares the performances of the algorithm with different matching functions when applied to the first 10 frames of the FOREMAN sequence at a frame rate of 8.33 frames/s (i.e., a frame skip[4] of 3). The quoted PSNR values are for the luma component only. It can be seen from this figure that the SSD measure achieves the best performance, followed very closely by the SAD measure. The NCCF measure, on the other hand, has the worst performance. While Figure 4.3 compares the performance in terms of prediction quality, Table 4.1 compares the performances in terms of computational complexity. It can be seen that the SAD measure has the lowest computational complexity, because it involves no multiplications. Because of its good prediction quality and small computational complexity, SAD is preferred by most implementations. All subsequent results assume the use of SAD as the matching function.

There are many other proposed matching functions. Most of them attempt to further reduce complexity, but this is often at the expense of a reduced prediction quality. A more detailed discussion of such functions is deferred to Chapter 7.

---

[4]Throughout this book, the term *frame skip* will be used to quantify the amount of temporal subsampling with respect to the original frame rate. For example, a frame skip of 3 means that the original sequence is temporally subsampled by a factor of 3:1. Thus, if the original sequence has a frame rate of 30 frames/s, then the subsampled sequence will have a frame rate of $30/3 = 10$ frames/s.

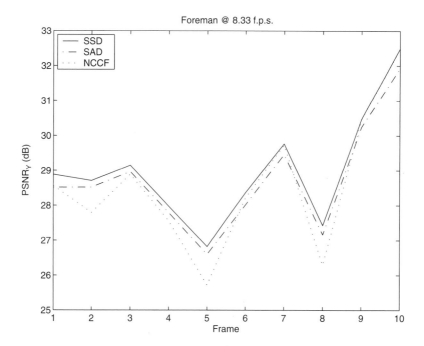

Figure 4.3: Reconstruction quality of SSD, SAD, and NCCF

Table 4.1: Computational complexity of SSD, SAD, and NCCF for an $N \times N$ block

	SAD	SSD	NCCF
$\mid \cdot \mid$	N^2	–	–
–	N^2	N^2	-
+	$N^2 - 1$	$N^2 - 1$	$3(N^2 - 1)$
×	–	N^2	$3N^2 + 1$
÷	–	–	1
$\sqrt{\ }$	–	–	2

4.6.2 Block Size

Another important parameter of the BMA is the block size. Figure 4.4 shows the performance of the BMA with two different sizes, 8×8 and 16×16. It can be seen in Figure 4.4(a) that a smaller block size achieves better prediction quality. This is due to a number of reasons. A smaller block size reduces the effect of the accuracy problem. In other words, with a smaller block size, there is less possibility that the block will contain different objects moving in

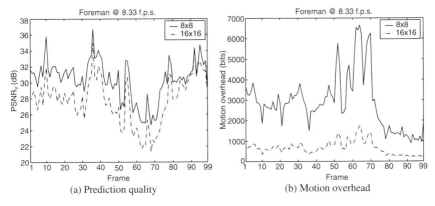

Figure 4.4: Performance of the BMA with different block sizes

different directions. In addition, a smaller block size provides a better piecewise translational approximation to nontranslational motion. Since a smaller block size means that there are more blocks (and consequently more motion vectors) per frame, this improved prediction quality comes at the expense of a larger motion overhead, as can be seen in Figure 4.4(b). Most video coding standards use a block size of 16×16 as a compromise between prediction quality and motion overhead. A number of variable-block-size motion estimation methods have also been proposed in the literature [98, 99]. As already discussed, the advanced prediction mode of the H.263 standard allows adaptive switching between block sizes of 16×16 and 8×8 on an MB basis.

4.6.3 Search Range

The maximum allowed motion displacement d_m, also known as the *search range*, has a direct impact on both the computational complexity and the prediction quality of the BMA. A small d_m results in poor compensation for fast-moving areas and consequently poor prediction quality. This is evident from Figure 4.5(a), which compares the performance of two ranges, ± 5 and ± 15. A large d_m, on the other hand, results in better prediction quality but leads to an increase in the computational complexity (since there are $(2d_m+1)^2$ possible blocks to be matched in the search window). A larger d_m can also result in longer motion vectors and consequently a slight increase in motion overhead,[5] as can be seen from Figure 4.5(b). In general, a maximum allowed

---

[5]As will be shown later, in block-motion fields, larger displacements are, in general, less probable. Thus, most video codecs assign longer codewords for longer motion vectors.

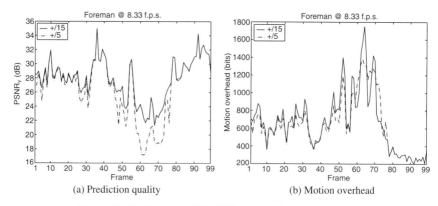

(a) Prediction quality   (b) Motion overhead

Figure 4.5: Performance of the BMA with different search ranges

displacement of $d_m = \pm 15$ pels is sufficient for low-bit-rate applications. As already discussed, the H.263 standard uses a maximum displacement of about ± 15 pels, although this range can optionally be doubled with the unrestricted motion vector mode.

4.6.4 Search Accuracy

Initially, the BMA was designed to estimate motion displacements with full-pel accuracy. Clearly, this limits the performance of the algorithm, since in reality the motion of objects is completely unrelated to the sampling grid. A number of workers in the field have proposed to extend the BMA to subpel accuracy. For example, Ericsson [100] demonstrated that a prediction gain of about 2 dB can be obtained by moving from full-pel to 1/8-pel accuracy. Girod [92] presented an elegant theoretical analysis of motion-compensating prediction with subpel accuracy. He termed the resulting prediction gain the *accuracy effect*. He also showed that there is a "critical accuracy" beyond which the possibility of further improving prediction is very small. He concluded that with block sizes of 16×16, quarter-pel accuracy is desirable for broadcast TV signals, whereas half-pel accuracy appears to be sufficient for videophone signals. Today, most video coding standards adopt subpel accuracy in its half-pel form. In fact, it has been shown [65] that most of the performance gain of H.263 over H.261 can be attributed to the move from full-pel to half-pel accuracy.

It should be pointed out, however, that the improved prediction quality of subpel accuracy comes at the expense of a significant increase in computational complexity. This increase is due to two reasons. First, the reference frame intensities have to be interpolated at subpel locations. Second, there are now

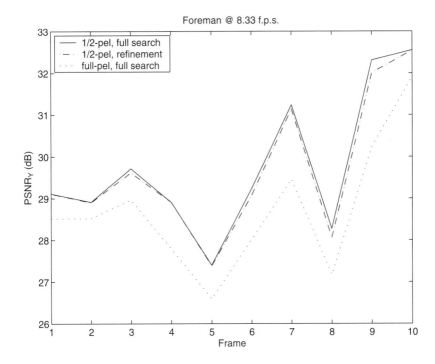

Figure 4.6: Performance of the BMA with subpel accuracy

more possible candidate blocks within the search window. For example, when moving from full-pel to half-pel accuracy, the number of candidate blocks in the search window increases from $(2d_m + 1)^2$ to $(4d_m + 1)^2$. To alleviate this complexity, most video codecs implement subpel accuracy as a postprocessing stage, where first a full-pel motion vector is obtained, usually using full search, and then this vector is *refined* to subpel accuracy using a limited search. This provides a large saving in computational complexity and at the same time maintains the improved prediction quality, as can be seen in Figure 4.6.

4.6.5   Unrestricted Motion Vectors

In some cases (like, for example, in border blocks) part of the search window is outside the reference frame area. This means that some of the candidate blocks in the search window are either partially or completely out of the reference frame. There are two ways to handle such candidate blocks. In the *restricted motion vectors* method, such blocks are ignored and skipped during motion estimation. In the *unrestricted motion vectors* method, however, such

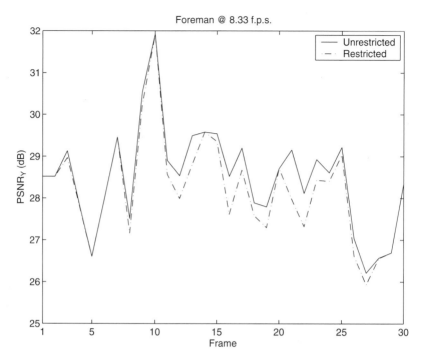

Figure 4.7: Performance of the BMA with restricted and unrestricted motion vectors

blocks are included in the motion estimation and compensation process. In this case, a referenced pel outside the frame is usually approximated by the closest border pel. This unrestricted method can improve the prediction quality along frame borders, especially in cases of camera or background movement. This is particularly useful in small frame formats, where border blocks represent a high percentage of the frame area. Figure 4.7 illustrates this improvement for part of the FOREMAN sequence. The method is included in the H.263 optional unrestricted motion vector mode and also in the advanced prediction mode.

4.6.6   Overlapped Motion Compensation

As already discussed, the BMA assumes that each block of pels moves with a uniform translational motion. Because this assumption does not always hold true, the method is known to produce blocking artefacts in the reconstructed frames. One method that reduces this effect is *overlapped motion compensation* (OMC). The method was first proposed by Watanabe and Singhal in 1991 [101]. In BMA, the estimated block motion vector is used to copy a displaced

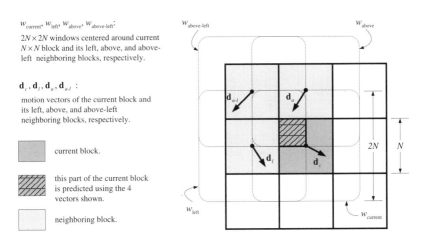

$W_{current}$, W_{left}, W_{above}, $W_{above-left}$:

$2N \times 2N$ windows centered around current $N \times N$ block and its left, above, and above-left neighboring blocks, respectively.

$\mathbf{d}_c$, $\mathbf{d}_l$, $\mathbf{d}_a$, $\mathbf{d}_{a-l}$:

motion vectors of the current block and its left, above, and above-left neighboring blocks, respectively.

current block.

this part of the current block is predicted using the 4 vectors shown.

neighboring block.

Figure 4.8: Overlapped motion compensation for the top-left quadrant of the current block

$N \times N$ block from the reference frame to the current $N \times N$ block in the current frame. In OMC, however, the estimated block motion vector is used to copy a larger block (say, $2N \times 2N$) from the reference frame to a position centered around the current $N \times N$ block. As illustrated in Figure 4.8, since they are larger than the compensated blocks, the copied blocks overlap, hence the name *overlapped motion compensation*. Each copied block is weighted by a smooth window, with higher weights at the center and lower weights toward the borders. This means that the estimated motion vector is given more influence in the center of the block, and this influence decays toward the borders, where neighboring motion vectors start taking over. This ensures a smooth transition between blocks and therefore reduces blocking artefacts. Overlapped motion estimation and compensation can also be implemented in the frequency domain, as proposed by Young and Kingsbury [95].

Another view of the OMC process is that each pel in the current $N \times N$ block is compensated using more than one motion vector. For example, in Figure 4.8, each pel is compensated using four motion vectors. The set of motion vectors is decided according to the spatial position of the pel within the block. A pel in the top-left quadrant of the current block will be compensated using the motion vector of the block itself, plus the motion vectors of the blocks to the left of, above, and above left of the current block. Each vector provides a prediction for the pel, and those four predictions are weighted according to the spatial position of the pel within the block. For example, as the spatial position of the pel gets closer to the left border of the block, a higher weight is given to the prediction provided by the motion vector of the block to the left.

Orchard *et al.* [102, 103] used this view to formulate OMC as a linear estimator of the form

$$\hat{f}_t(\mathbf{s}) = \sum_{\mathbf{d}_n \in \mathcal{N}(\mathbf{s})} w_n(\mathbf{s}) \, f_{t-\Delta t}(\mathbf{s} - \mathbf{d}_n), \tag{4.32}$$

where $\mathcal{N}(\mathbf{s}) = \{\mathbf{d}_n(\mathbf{s})\}$ is the set of motion vectors used to compensate the pel at location $\mathbf{s}$ and $w_n(\mathbf{s})$ is the weight given to the prediction provided by vector $\mathbf{d}_n$. Using this formulation, they solve two optimization problems: overlapped-motion compensation and overlapped-motion estimation. Given the set of motion vectors $\mathcal{N}(\mathbf{s})$ estimated by the encoder, they propose a method for designing optimal windows, $w_n(\mathbf{s})$, to be used at the decoder for motion compensation. Also, given a fixed window that will be used at the decoder, they propose a method for finding the optimal set of motion vectors at the encoder. Note that the latter problem is much more complex than the BMA, since in this case the estimated motion vectors are interdependent. For this reason, their proposed method is based on an iterative procedure. A number of methods have been proposed to alleviate this complexity, e.g., Ref. 104.

As a linear estimator of intensities, OMC belongs to a more general set of motion compensation methods called *multihypothesis motion compensation*. Another member in this set is bidirectional motion compensation. The theoretical motivations for such methods were presented by Sullivan in 1993 [105]. Recently, Girod [106] analyzed the rate-distortion efficiency of such methods and provided performance bounds and comparisons with single-hypothesis motion compensation (e.g., the BMA).

Figure 4.9 compares the performance of OMC to that of the BMA when applied to the FOREMAN sequence. In the case of OMC, the same BMA motion vectors were used for compensation (i.e., the motion vectors were not optimized for overlapped compensation). Each motion vector was used to copy a 32×32 block from the reference frame and center it around the current 16×16 block in the current frame. Each copied block was weighted by a bilinear window function defined as [103]

$$w(x, y) = w_x \cdot w_y, \quad \text{where } w_z = \begin{cases} \frac{1}{16}(z + \frac{1}{2}) & \text{for } z = 0, \ldots, 15, \\ w_{31-z} & \text{for } z = 16, \ldots, 31. \end{cases} \tag{4.33}$$

Border blocks were handled by assuming "phantom" blocks outside the frame boundary with motion vectors equal to those of the border blocks. Despite the fact that the estimated vectors, the window shape, and the overlapping weights were not optimized for overlapped compensation, OMC provided better objective (Figure 4.9(a)) and subjective (Figures 4.9(b)–4.9(d)) quality compared to the BMA. In particular, the annoying blocking artefacts have clearly been reduced.

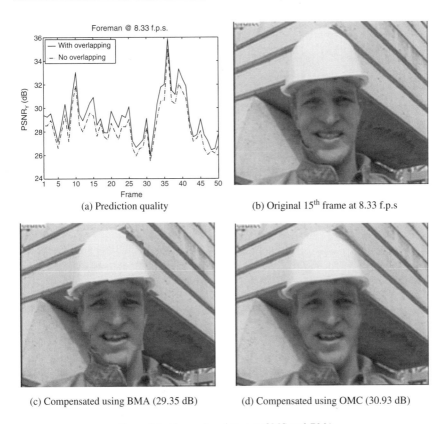

(a) Prediction quality

(b) Original 15th frame at 8.33 f.p.s

(c) Compensated using BMA (29.35 dB)

(d) Compensated using OMC (30.93 dB)

Figure 4.9: Comparison between OMC and BMA

4.6.7 Properties of Block-Motion Fields and Error Surfaces

This subsection presents some basic properties of the BMME algorithm when applied to typical video sequences. These properties will be utilized and referenced in subsequent chapters of the book. All illustrations in this subsection were generated using a full-pel full-search block-matching algorithm with 16×16 blocks, ± 15 pels maximum displacement, restricted motion vectors, SAD as the BDM, and original reference frames.

Property 4.6.7.1 *The distribution of the block motion field is center-biased.* This means that smaller displacements are more probable and the motion vector $(0, 0)$ has the highest probability of occurrence. In other words, most blocks are stationary or quasi-stationary. This property is illustrated in Figure 4.10(a) for AKIYO at 30 frames/s (frame skip of 1). The property also holds true for

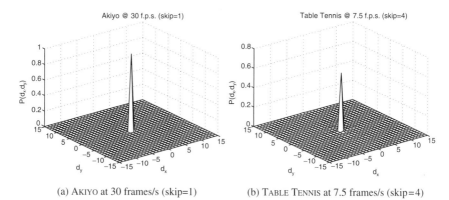

(a) AKIYO at 30 frames/s (skip=1)   (b) TABLE TENNIS at 7.5 frames/s (skip=4)

Figure 4.10: Center-biased distribution of block-motion field

Foreman at 25 frames/s		
$\rho_x = 0.56$ $\rho_y = 0.33$	$\rho_x = 0.69$ $\rho_y = 0.49$	$\rho_x = 0.64$ $\rho_y = 0.46$
$\rho_x = 0.66$ $\rho_y = 0.48$	current block	$\rho_x = 0.72$ $\rho_y = 0.63$
$\rho_x = 0.64$ $\rho_y = 0.43$	$\rho_x = 0.76$ $\rho_y = 0.61$	$\rho_x = 0.68$ $\rho_y = 0.56$

(a) Correlation coefficients betwee the motion vector of a block and its eight neighboring blocks

(b) Distribution of the diffference between the horizontal component of the current vector and that of its left neighbor

Figure 4.11: Highly correlated block-motion fields

sequences with higher motion content and at lower frame rates, as illustrated in Figure 4.10(b) for TABLE TENNIS at 7.5 frames/s (frame skip of 4).

Property 4.6.7.2 *The block motion field is smooth and varies slowly.* In other words, there is high correlation between the motion vectors of adjacent blocks. Thus, it is very common to find neighboring blocks with identical or nearly identical motion vectors. This is evident in Figure 4.11(a), which shows the correlation coefficients between the motion vector of a block and its eight

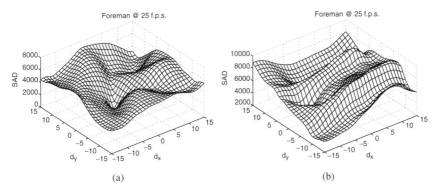

Figure 4.12: Sample multimodal error surfaces

neighboring blocks in FOREMAN at 25 frames/s. This is also illustrated in Figure 4.11(b), which shows the distribution of the difference between the horizontal component of the current vector (C_{d_x}) and that of its left neighbor (L_{d_x}). The bias of this distribution toward the zero difference clearly indicates high correlation, and this holds true for both AKIYO at 30 frames/s and TABLE TENNIS at 7.5 frames/s.

Property 4.6.7.3 *The error surface is usually multimodal.* In most cases, the error surface will contain one or more local minima, as illustrated in Figure 4.12. This can be due to a number of reasons, for example, the ambiguity problem, the accuracy problem, and the textured (periodical) local frame content.

Property 4.6.7.4 *The value of the global minimum of an error surface can change according to many factors, such as the frame skip, the motion content, and the block content.* For example, Figure 4.12 shows the error surface of two blocks from the same frame. The value of the global minimum of the surface in Figure 4.12(a) is 614, whereas that of the surface in Figure 4.12(b) is 3154.

4.7 A Comparative Study

This section presents the results of a comparative study of the motion estimation methods discussed in Sections 4.3–4.6. The main aim of this study is to answer the following question: *What is the best motion estimation algorithm for video coding?* In this study, the following algorithms were implemented:

DFA This is an implementation of the differential method of Cafforio and Rocca as given by Equation (4.12). In this case, the moving area, $\mathscr{A}$, was set to a block of 16×16 pels.

PRA This is an implementation of the *pel-recursive algorithm* of Netravali and Robbins as given by Equation (4.20). In this case, the motion vector of the previous pel in the line was taken as the initial motion estimate, $\hat{\mathbf{d}}^i$, of the current pel, the update step size was set to $\varepsilon = 1/1024$, the update term was calculated and averaged over an area of 3×3 pels centered around the current pel, and five iterations were performed per pel.

PCA This is an implementation of the phase correlation method as given by Equations (4.24) and (4.25). In this case, a window of 32×32 pels centered around the current 16×16 block was used to generate the phase correlation surface. The three most dominant peaks in this surface were detected and the corresponding motion displacements were unwrapped using Equation (4.27). The three candidate displacements were then tested using the SAD between the current block and the candidate displaced block in the reference frame. The candidate displacement with the lowest SAD was chosen as the motion vector of the current block.

BMA This is an implementation of a full-search block-matching algorithm. In this case, the block size was 16×16 pels and the matching criterion was the SAD.

In each case, the maximum allowed motion displacement was set to ± 15 pels in each direction and the motion vectors were allowed to point outside the reference frame (i.e., unrestricted motion vectors). To provide a fair comparison and to ease motion vector coding, all displacements were estimated with half-pel accuracy. In DFA and PRA this was achieved by rounding the subpel accurate motion estimates to the nearest half-pel accurate motion vectors. In PCA and BMA this was achieved using a refinement stage that examined the eight nearest half-pel estimates centered around the full-pel motion estimate. Bilinear interpolation was used to obtain intensity values at subpel locations of the reference frame. To mask the effect of the temporal propagation of prediction errors, motion was estimated and compensated using original reference frames. For comparison purposes, motion vectors were coded using the median predictor and the VLC table of the H.263 standard. The DFD signal was also transform encoded according to the H.263 standard and a quantization parameter of $QP = 10$. All quoted results refer to the luma components of sequences. No chroma encoding was performed.

Care should be taken when interpreting the results of this study. Different simulation parameters will lead to different results. For example, at the expense of a higher computational complexity, the performance of the PRA can be improved by increasing the number of iterations. This is also true when examining more peaks for the PCA.

Figure 4.13 compares the prediction quality of the four algorithms when applied to the three test sequences AKIYO, FOREMAN, and TABLE TENNIS, at different frame skips (and, consequently, different frame rates).

As expected, the DFA performs well for sequences with a low amount of movement (AKIYO) and at low frame skips (i.e., high frame rates). For sequences with a higher amount of movement (FOREMAN and TABLE TENNIS) and also at high frame skips, the motion vectors become longer, the quality of the Taylor series approximation becomes poor, and the performance of the DFA deteriorates.

Due to its dense motion field, the PRA has a superior performance for AKIYO and a very competitive performance for FOREMAN and TABLE TENNIS. The relative drop in performance for high-motion sequences and at high frame skips may be due to a number of reasons. With longer motion vectors, there is more possibility that the algorithm will be trapped in a local minimum before reaching the global minimum. Also, the maximum number of iterations may not be sufficient to reach the global minimum. However, increasing the number of iterations will increase the complexity of the algorithm.

In general, the performance of the PCA is somewhere in between that of the DFA and PRA. The poor performance for AKIYO may be due to the spurious peaks produced by the boundary and spectral leakage effects. Such effects may be reduced by applying a weighting function to smooth the phase correlation surface.

The best overall performance is provided by the BMA. It performs well regardless of the sequence type and the frame skip. In fact, for sequences with a high amount of movement (FOREMAN and TABLE TENNIS), the BMA shows superior performance.

It is interesting at this point to concentrate on the PRA and BMA, for two reasons. First, they achieved the best prediction quality performance in the comparison. Second, they represent two different approaches to motion estimation (pel-based and block-based, respectively). Figure 4.14 compares the performance of the PRA and the BMA for the first 50 frames of the FOREMAN sequence at 25 frames/s. Two versions of the PRA are considered: PRA, which is the same algorithm described earlier, and PRA-C, which is an algorithm in which the update term is based on the *causal* part of an area of 5×5 pels centered around the current pel. Since PRA-C is based on causal data, no motion overhead needs to be transmitted for this method. Due to the high amount of motion in FOREMAN, the maximum number of iterations for both pel-recursive algorithms was increased to 10.

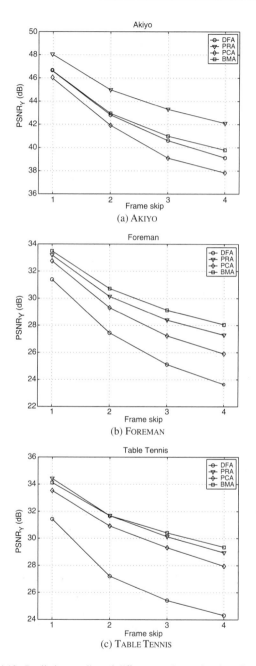

(a) AKIYO

(b) FOREMAN

(c) TABLE TENNIS

Figure 4.13: Prediction quality of different motion estimation algorithms

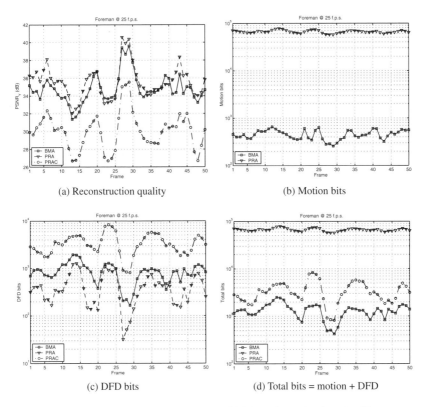

(a) Reconstruction quality   (b) Motion bits

(c) DFD bits   (d) Total bits = motion + DFD

Figure 4.14: Comparison between BMA and PRA motion estimation algorithms

The aim of motion estimation for video coding is to simultaneously minimize the bit rate corresponding both to the motion parameters (motion bits) and to the prediction error signal (DFD bits). As illustrated in Figure 4.14, the three algorithms represent three different tradeoffs between prediction quality and motion overhead. Due to its dense motion field, the PRA has the best prediction quality and, consequently, the least DFD bits. This is, however, at the expense of a prohibitive motion overhead, which leads to a very high total bit rate. The causal implementation of the PRA, PRA-C, clearly restricts the method and significantly reduces its prediction quality. Thus, PRA-C removes the motion overhead at the expense of an increase in DFD bits. In addition, this causal implementation increases the complexity of the decoder. The best tradeoff is achieved by the BMA. It uses a block-based approach to reduce the motion overhead while still maintaining a very good prediction quality. This explains the popularity of this approach and its inclusion in video coding standards.

4.8   Efficiency of Block Matching at Very Low Bit Rates

The incorporation of motion estimation and compensation into a video codec involves extra computational complexity. This extra complexity must, therefore, be justified on the basis of an enhanced coding efficiency. This is very important for very-low-bit-rate applications and, in particular, for applications like mobile video communication, where battery time and processing power are scarce resources.

Very low bit rates are usually associated with high frame skips. As the frame skip increases, the temporal correlation between consecutive frames decreases. This will obviously decrease the efficiency of motion estimation, as can be seen in Figure 4.13. This poses a very important question: Is the use of motion estimation at such bit rates justifiable? Or put in another way, Is the use of less complex coding methods, like frame differencing and intraframe coding, sufficient at those bit rates?

This study investigates the efficiency of block-matching motion estimation at very low bit rates. Three algorithms were implemented:

BMA-H This is a half-pel full-search BMA with 16×16 blocks, ± 15 pels maximum displacement, restricted motion vectors, and SAD as the matching criterion. Half-pel accuracy is achieved using a refinement stage around the full-search full-pel motion vectors. Bilinear interpolation is used to obtain intensity values at subpel locations of the reference frame.

FDIFF This is a frame differencing algorithm. This means that no motion estimation is performed and the motion vectors are always assumed to be $(0, 0)$. Note that this algorithm has no motion overhead and the total frame bits are equal to the DFD bits.

INTRA This is a DCT-based intraframe coding algorithm.

In each algorithm, motion was estimated and compensated using reconstructed reference frames. Motion vectors were coded using the median predictor and the VLC table of the H.263 standard. Both, the DFD signal (in the case of BMA-H and FDIFF) and the frame signal (in the case of INTRA) were transform encoded according to the H.263 standard. To simulate a very-low-bit-rate environment, the frame skip was set to 4 (this corresponds to 7.5 frames/s for AKIYO and TABLE TENNIS and to 6.25 frames/s for FOREMAN). To generate a range of bit rates, the quantization parameter QP was varied over the range 5–30 in steps of 5. This means that each algorithm was used to encode a given sequence six times. Each time, QP was held constant over the whole sequence (i.e., no rate control was used). The first frame of a sequence was always INTRA coded, regardless of the encoding algorithm, and the

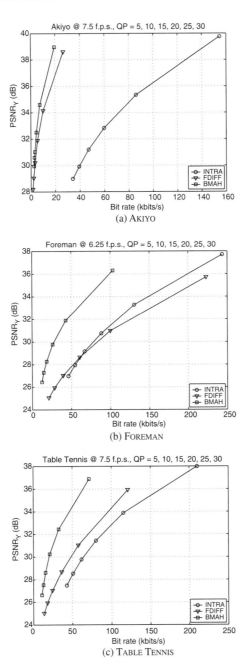

Figure 4.15: Efficiency of block-matching motion estimation at very low bit rates

resulting bits were included in the bit-rate calculations. All quoted results refer to the luma components of sequences. Figure 4.15 compares the performance of the three algorithms when applied to the three test sequences.

In general, both interframe coding algorithms (FDIFF and BMA-H) outperform the intraframe coding algorithm (INTRA). Thus, even at very low bit rates, high frame skips, or low-motion sequences, the temporal correlation between video frames is still high enough to justify interframe coding.

Comparing the two interframe coding algorithms, it is immediately evident that the BMA-H algorithm outperforms the FDIFF algorithm at all bit rates and for all sequences. Note, however, that at extremely low bit rates, and in particular for the low-motion AKIYO sequence, the efficiency of the BMA-H algorithm starts to drop and approaches that of the simpler FDIFF algorithm. But even with this drop in performance, the use of BMA-H is still justifiable. For example, with AKIYO and at a bit rate as low as 3 kbits/s, the BMA-H algorithm still outperforms the FDIFF algorithm by about 1 dB.

4.9 Discussion

Motion estimation is an important process in a wide range of applications. Different applications have different requirements and may, therefore, employ different motion estimation techniques.

In video coding, the determination of the *true* motion is not the intrinsic goal. The aim is rather to simultaneously minimize the bit rate corresponding both to the motion parameters (motion bits) and to the prediction error signal (DFD bits). This is not an easy task, since the minimization of one quantity usually leads to maximizing the other. Thus, a suitable tradeoff is usually sought. In this chapter, four motion estimation methods were compared. The four methods are the differential, pel-recursive, phase-correlation, and block-matching motion estimation methods. It was found that block-matching motion estimation provides the best tradeoff. It uses a block-based approach to reduce the motion overhead while still maintaining a very good prediction quality (and consequently a small number of DFD bits). This explains the popularity of this approach and its inclusion in video coding standards.

The chapter also investigated the efficiency of motion estimation at very low bit rates. It was found that the prediction quality of motion estimation starts to drop at very low bit rates, in particular, for low-motion sequences, and approaches that of simpler techniques, like frame differencing and intraframe coding. Despite this drop in prediction quality, it was found that the use of motion estimation is still justifiable at those bit rates.

Chapter 5

Warping-Based Motion Estimation Techniques

5.1 Overview

As already discussed, one way to achieve higher coding efficiency is to improve the performance of the motion estimation and compensation processes. This can be done by using advanced motion estimation and compensation techniques. This chapter concentrates on an advanced technique called *warping-based motion estimation*. Since the early 1990s, this technique has attracted attention in the video coding community as an alternative to (or rather as a generalization of) conventional block-matching methods.

Section 5.2 reviews warping-based motion estimation techniques. Various aspects of such techniques, like the shape of patches, the type of meshes, the spatial transformation, the continuity of the motion field, the direction of node tracking, the node-tracking algorithm, the motion compensation method, and the transmitted motion overhead, are considered and compared. Section 5.3 compares the performance of warping-based methods to that of block-matching methods. In particular, the section investigates the efficiency of warping-based methods at very low bit rates. The chapter concludes with a discussion in Section 5.4.

5.2 Warping-Based Methods: A Review

Motion estimation (ME) can be defined as a process that divides the current frame, f_c, into regions and that estimates for each region a set of *motion parameters*, $\{a_i\}$, according to a motion model. The *motion compensation* (MC) process then uses the estimated motion parameters and the motion model to *synthesize* a prediction, $\hat{f_c}$, of the current frame from a reference frame, f_r.

This synthesis process can be formulated as follows:

$$\hat{f_c}(x, y) = f_r(u, v), \tag{5.1}$$

where (x, y) are the spatial coordinates in the current frame (or its prediction) and (u, v) are the spatial coordinates in the reference frame. This equation indicates that the MC process applies a *geometric transformation* that maps one coordinate system onto another. This is defined by means of the *spatial transformation* functions g_x and g_y:

$$\begin{aligned} u &= g_x(x, y), \\ v &= g_y(x, y). \end{aligned} \tag{5.2}$$

This spatial transformation is also referred to as *texture mapping* or *image warping* [107].

As already discussed, the BMA method relies on a uniform translational motion model. Thus, the transformation functions of this method are given by

$$\begin{aligned} u &= g_x(x, y) = x + a_1 = x + d_x, \\ v &= g_y(x, y) = y + a_2 = y + d_y. \end{aligned} \tag{5.3}$$

In practice, however, a block can contain multiple moving objects, and the motion is usually more complex and can contain translation, rotation, shear, expansion, and other deformation components. In such cases, the simple uniform translational model will fail, and this will usually appear as artefacts, e.g., blockiness, in the motion-compensated prediction. Higher-order motion models can be used to overcome such problems. Examples of such models are the affine, bilinear, and perspective spatial transformations given by Equations (5.4), (5.5), and (5.6), respectively:

Affine:

$$\begin{aligned} u &= g_x(x, y) = a_1 x + a_2 y + a_3, \\ v &= g_y(x, y) = a_4 x + a_5 y + a_6. \end{aligned} \tag{5.4}$$

Bilinear:

$$\begin{aligned} u &= g_x(x, y) = a_1 xy + a_2 x + a_3 y + a_4, \\ v &= g_y(x, y) = a_5 xy + a_6 x + a_7 y + a_8. \end{aligned} \tag{5.5}$$

Perspective:

$$u = g_x(x, y) = \frac{a_1 x + a_2 y + a_3}{a_7 x + a_8 y + 1},$$

$$v = g_y(x, y) = \frac{a_4 x + a_5 y + a_6}{a_7 x + a_8 y + 1}.$$

(5.6)

Motion estimation and compensation using higher-order models is usually performed using the following steps:

1. A 2-D *mesh* is used to divide the current frame into nonoverlapping polygonal patches (or elements). The points shared by the vertices of the patches are referred to as *grid* or *node points*.

2. The motion of each node is estimated. This will map each node in the current frame to a corresponding node in the reference frame. In effect, this will map each patch in the current frame to a corresponding patch in the reference frame.

3. For each patch in the current frame, the coordinates of its vertices and those of the matching patch in the reference frame are used to find the motion parameters $\{a_i\}$ of the underlying motion model.

4. During motion compensation, the estimated motion parameters $\{a_i\}$ are substituted in the appropriate spatial transformation, Equations (5.4)–(5.6), to *warp* the patch in the reference frame to provide a prediction for the corresponding patch in the current frame.

An example of this process is illustrated in Figure 5.1. In this figure the current frame is divided into square patches. This forms a uniform mesh. During motion estimation, node points A, B, C, and D in the current frame are mapped to node points A', B', C', and D' in the reference frame. During motion compensation, the deformed patch A'B'C'D' is warped to provide a prediction for the square patch ABCD.

It should be pointed out that there is a lack of consistency in the literature when referring to this type of motion estimation and compensation methods. Examples of the numerous names employed are control grid interpolation [108, 109, 110], warping-based methods [111, 112, 113], spatial-transformation-based methods [114, 115, 116, 117], geometric-transformation-based methods [118], generalized motion estimation methods [119, 120], and mesh-based methods [121, 122, 123, 124, 125, 126, 127].

When designing a warping-based technique, several aspects of the method need to be considered and defined, as discussed in the following subsections.

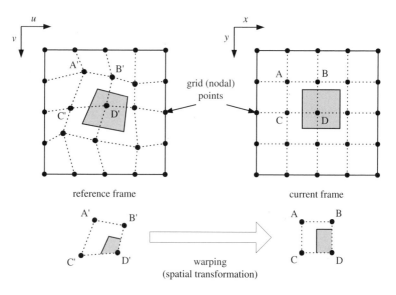

Figure 5.1: Warping-based motion estimation and compensation

5.2.1   Shape of Patches

The most widely used shapes are *triangles* and *quadrilaterals*. Nakaya and Harashima [114] showed that equilateral triangles are optimal, in the prediction-quality sense, when the affine transformation is used, whereas squares are optimal when the bilinear transformation is used. Square patches are some-times preferred because they are compatible with current block-based video coding methods and standards. Triangular patches are more compatible with model-based coding methods, where wireframe models are usually defined in terms of triangles.

5.2.2   Type of Mesh

The mesh structure can be fixed or adaptive. A *fixed mesh* is one that is built according to a predetermined pattern, e.g., a regular mesh with square patches. An *adaptive mesh*, on the other hand, is one that is adaptively built according to frame contents and motion. Adaptive meshes can be content-based or motion-based. In content-based adaptive meshes, nodes are placed to fit important features like contours and edges [111, 121]. In motion-based adaptive meshes, more nodes are placed in moving areas. This is usually achieved using a hierarchical (usually, quad-tree) mesh structure [109, 120,

115, 123]. Although adaptive meshes can improve prediction quality, they have the disadvantages of increased computational complexity (for the generation and adaptation processes) and increased overhead (to describe the structure of the mesh). The structure overhead can be removed by applying the adaptation process based on previous frames that are available at the decoder.

5.2.3 Spatial Transformation

As shown by Seferidis and Ghanbari [119], the perspective transform achieves the best prediction-quality performance. However, the high computational complexity of this transformation limits its use in practice. The affine transformation is the least computationally complex, but it has the fewest degrees of freedom. The performance of the bilinear transformation is very close to that of the perspective transformation, with the advantage of reduced computational complexity. However, a study by Nakaya and Harashima [114] showed that the affine and bilinear transformations have almost the same performance when the patch shape is optimized (equilateral triangles and squares, respectively). In fact, the same study showed that the performance of the affine transformation can be superior as the number of nodes decreases.

5.2.4 Continuous Versus Discontinuous Methods

Adjacent patches in the current frame have common vertices between them. There are two main methods for estimating the motion of such common vertices. If the motion of common vertices is estimated independent from each other (i.e., common vertices are assigned different motion vectors), then this will result in a discontinuous motion field with discontinuities along the boundaries of the patches. This is known as the *discontinuous method*. The motion field in this case has similarities with that produced by the BMA. If, however, a restriction is applied such that common vertices have the same motion vector, then this will result in a continuous motion field and the method is known as the *continuous method*. The two methods are illustrated in Figure 5.2.

As pointed out by Ghanbari *et al.* [115], the discontinuous method is more flexible and can compensate for more general complex motion. However, as pointed out by Nakaya and Harashima [114], since discontinuities are allowed along the boundaries of patches, this method can suffer from blocking artefacts. Another disadvantage of the discontinuous method is that it generates more motion overhead (four motion vectors per patch) compared to the continuous method (about one motion vector per patch).

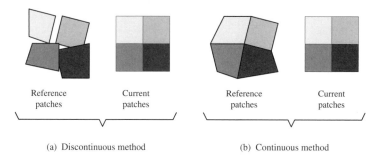

(a) Discontinuous method   (b) Continuous method

Figure 5.2: Continuous versus discontinuous warping-based methods

5.2.5 Backward Versus Forward Node Tracking

The process of estimating the motion of a grid or a node point is called *node tracking*. There are two types of node-tracking algorithms: backward and forward node tracking.

In *backward node tracking*, nodes are first placed on the current frame and then they are matched to points in the reference frame. During motion compensation, a pel (x, y) in the current patch is *copied from* a corresponding pel $(u, v) = (g_x(x, y), g_y(x, y))$ in the reference patch. Note that in this case, (x, y) is a sampling spatial position, whereas (u, v) may be a nonsampling spatial position. Interpolation, e.g., bilinear, can be used to obtain pel values at nonsampling positions of the reference frame. This process is repeated for all pels within the current patch. Since backward tracking starts with a mesh on the current frame (which is not available at the decoder), this technique is usually used in combination with a fixed mesh.

In *forward node tracking*, nodes are first placed on the reference frame and then matched to points in the current frame. During motion compensation, a pel (u, v) in the reference patch is *copied to* a corresponding pel $(x, y) = (g_x(u, v), g_y(u, v))$ in the current patch. Since, in this case, (x, y) may be a nonsampling spatial position, the compensated current patch will normally contain holes (i.e., noncompensated pels at sampling spatial positions). Techniques that can be used to recover pel values at sampling spatial positions from values at nonsampling spatial positions are discussed and compared by Sharaf and Marvasti in Ref. 116. Due to the use of such techniques, forward node tracking and compensation is computationally more complex than backward node tracking and compensation. Since forward node tracking starts with a mesh on the reference frame, this technique is usually used in combination with an adaptive mesh. Although the combination of forward tracking and adaptive meshes can provide some prediction-quality improvement over the combination of backward tracking and fixed meshes, the use of the former is not justified, due to the huge increase in computational complexity [116].

5.2.6 Node-Tracking Algorithm

A simple method to estimate the motion of a node is to use a BMA-type algorithm which minimizes the translational prediction error in a block centered around the node. Niewęgłowski *et al.* [111] use a modified BMA with a large block (21×21) centered around the node and a distortion measure designed to give more weight to pels closer to the node. To reduce complexity, the block is subsampled by a factor of 2:1 in both directions.

Although BMA-type algorithms are simple, they provide suboptimal performance. First, they assume that the motion of a node is independent of the motion of other nodes, and second, they assume that minimizing the translational prediction error minimizes the true prediction error. In practice, however, both assumptions are not true. A node is a common vertex between more than one patch. Consequently, the displacement of a node will affect all patches connected to it. For example, with quadrilateral patches, the displacement of a node affects the prediction quality within four patches connected to it. It follows that the choice of the motion vector of one node will affect the choice of the motion vectors of other nodes. In addition, the true prediction error is the error between the current frame and its warped prediction from the reference frame. This is not equal to the translational prediction error.

Brusewitz [128] uses a BMA-type algorithm to provide coarse approximations for nodal motion vectors. An iterative gradient-based approach that minimizes the true prediction error is then used to refine all nodal motion vectors simultaneously. The computational complexity of the method is extremely high. For example, if there are 100 nodes in the frame, the method requires the inversion of a 200×200 matrix.

To reduce complexity, Sullivan and Baker [108] estimate the motion of one node at a time. However, to take into account the interdependence between motion vectors, an iterative approach is employed. In each iteration, the nodes are processed sequentially. The motion vector of a node is estimated using a local search around the motion vector from the previous iteration while holding constant the motion vectors of its surrounding nodes. During the local search, the quality of a candidate motion vector is measured by calculating the distortion measure between all patches connected to the node and their warped predictions from the reference frame. The local search is applied to a node only if its motion vector, or the motion vector of at least one of its surrounding nodes, was changed in the previous iteration.

Nakaya and Harashima [114] use a *hexagonal matching algorithm* (HMA). The name is due to the use of triangular patches for which each node is a common vertex between six patches forming a hexagon. The algorithm is almost identical to that of Sullivan and Baker (described earlier). In this case, however, a BMA-type algorithm is first used to provide a coarse estimate of the motion field, and the iterative approach is then used to refine this estimate. In addition to the exhaustive local search, they also propose a faster but suboptimal gradient-based local search. Similar gradient-based approaches have also been used by Wang *et al.* [123, 126] and Dudon *et al.* [124].

Altunbasak and Tekalp [125] first estimate a dense field of motion displacements. Then they use a least squares method to estimate the nodal motion vectors subject to the constraint of preserving the connectivity of the mesh. They show that the performance of this algorithm is comparable to that of HMA, with the advantage of reduced computational complexity.

When estimating a nodal motion vector, it is very important to ensure that the estimate does not cause any patch connected to the node to become degenerate (i.e., with obtuse angles and/or flipover nodes). To accomplish this, Wang *et al.* [123, 126] limit the search range to a diamond region defined by the four surrounding nodes, whereas Altunbasak and Tekalp [125] use a postprocessing stage where an invalid estimate is replaced by a valid estimate interpolated from surrounding nodal motion vectors.

All the foregoing algorithms assume a continuous motion field. Ghanbari *et al.* [119, 120, 115, 117] use quadrilateral patches with a discontinuous motion field. In this case, the four vertices of each regular patch in the current frame are displaced combinatorially (i.e., perturbed) to find the best-match deformed patch in the reference frame. The computational complexity of this algorithm is extremely high since there are $(2d_m + 1)^8$ possible deformed patches in the reference frame. In addition, each possible patch must first be warped to calculate the distortion measure. To reduce complexity, they propose to use a fast-search algorithm, e.g. Ref. 129.

5.2.7   Motion Compensation Method

Having obtained nodal motion vectors, there are two methods of performing motion compensation.

In the first method, for each patch in the current frame, the coordinates of its vertices and those of the matching patch in the reference frame are used to set up a number of simultaneous equations. This set is then solved for the motion parameters $\{a_i\}$ of the underlying motion model. For example, assume a mesh of quadrilateral patches and a bilinear motion model. If the spatial coordinates of the top-left, top-right, bottom-left, and bottom-right vertices of the patch in the current frame are (x_A, y_A), (x_B, y_B), (x_C, y_C), and (x_D, y_D), respectively, and the corresponding estimated motion vectors are $\mathbf{d}_A$, $\mathbf{d}_B$, $\mathbf{d}_C$, and $\mathbf{d}_D$, respectively, then the spatial coordinates of the matching vertices in the reference frame are (u_A, v_A), (u_B, v_B), (u_C, v_C), and (u_D, v_D), respectively, where, e.g., $(u_A, v_A) = (x_A + d_{x_A}, y_A + d_{y_A})$. Using the bilinear model of Equation (5.5), the following set of simultaneous equations is obtained:

$$
\begin{bmatrix} u_A & u_B & u_C & u_D \\ v_A & v_B & v_C & v_D \end{bmatrix} = \begin{bmatrix} a_1 & a_2 & a_3 & a_4 \\ a_5 & a_6 & a_7 & a_8 \end{bmatrix} \cdot \begin{bmatrix} x_A y_A & x_B y_B & x_C y_C & x_D y_D \\ x_A & x_B & x_C & x_D \\ y_A & y_B & y_C & y_D \\ 1 & 1 & 1 & 1 \end{bmatrix}.
$$

$$(5.7)$$

This set can easily be solved for the motion parameters $a_1, \ldots, a_8$. Having obtained the motion parameters of the current patch, each pel (x, y) in the patch is then compensated from a pel (u, v) in the reference patch, where (u, v) are obtained using Equation (5.5).

In the second method of motion compensation (commonly known as *control grid interpolation* (CGI) [108]), the motion vectors at the vertices of the current patch are interpolated to produce a dense motion field within the patch. For the same example just given, the motion vector $\mathbf{d}(x, y) = (d_x(x, y), d_y(x, y))$ at pel (x, y) of the current patch is obtained by bilinear interpolation of the four motion vectors at the vertices. Thus

$$\mathbf{d}(x, y) = (1 - x_n)(1 - y_n)\mathbf{d}_A + x_n(1 - y_n)\mathbf{d}_B + (1 - x_n)y_n\mathbf{d}_C + x_n y_n \mathbf{d}_D,$$

$$(5.8)$$

$$\text{where} \quad x_n = \frac{x - x_A}{x_B - x_A} \quad \text{and} \quad y_n = \frac{y - y_A}{y_C - y_A}.$$

$$(5.9)$$

Each pel (x, y) in the current patch can then be compensated from pel (u, v) in the reference patch, where $(u, v) = (x + d_x(x, y), y + d_y(x, y))$. It can be shown [110] that the two methods are equivalent.

5.2.8 Transmitted Motion Overhead

Two types of motion overhead can be transmitted: the motion parameters a_i of the patches and the motion vectors of the nodes. Motion vectors have a limited range and are usually evaluated to a finite accuracy (e.g., full- or half-pel accuracy), whereas motion parameters are not limited and are usually continuous in value. Thus, motion vectors are usually preferred because they are easier to encode and result in a more compact representation. In addition, motion vectors ensure compatibility with current video coding standards. One disadvantage in this case, however, is that the decoder is more complex, since it must use the received motion vectors to calculate the motion parameters or to interpolate the motion field (as described in Section 5.2.7) before being able to perform motion compensation.

5.3 Efficiency of Warping-Based Methods at Very Low Bit Rates

This section investigates the performance of warping-based methods and compares it to that of block-matching methods. The main aim is to answer the following question: *Are there any gains for using higher-order motion models at very low bit rates?* In other words, this section assesses the suitability of warping-based methods for applications like mobile video communication.

Most results reported in the literature compare a warping-based algorithm to the basic block-matching algorithm. The authors feel that this is an unfair comparison for the following reasons:

1. As shown in Section 5.2.7, in warping-based compensation the motion vector used to compensate a pel in a given patch is interpolated from the nodal motion vectors at the vertices of the patch. Although the nodal motion vectors may be at full-pel accuracy, the resulting interpolated motion vector is at subpel accuracy. It is unfair to compare this subpel compensation to the full-pel compensation of the basic block-matching algorithm. A more fair comparison would be with a subpel (at least half-pel) block-matching algorithm.

2. Again, from Section 5.2.7, a warping-based method calculates one motion vector per pel. Thus, each pel within a patch is compensated individually. It is unfair to compare this to the basic block-matching algorithm, where the whole block is compensated using the same motion vector. A fairer comparison would be with overlapped motion compensation, where each pel within the block is compensated individually, as evident from Equation (4.32).

3. A warping-based method is much more computationally complex than the basic block-matching method (as is shown later). This increased complexity gives the warping-based method an unfair advantage over the basic block-matching method. To provide a fairer comparison, the basic block-matching method must be augmented by some advanced techniques (like subpel accuracy and overlapped compensation).

Thus, in this study, the following algorithms were implemented:

BMA This is a full-search full-pel block-matching algorithm with 16×16 blocks, restricted motion vectors, a maximum displacement of ± 15 pels, and SAD as the matching criterion.

BMA-HO This is the same as BMA but with half-pel accuracy and over-lapped motion compensation. Half-pel accuracy was obtained using a refinement stage around the full-pel motion vector. Overlapping windows of 32×32 and a bilinear weighting function, Equation (4.33), were used for overlapped motion compensation. Border blocks were handled by assuming "phantom" blocks outside the frame boundary, with motion vectors equal to those of the border blocks.

WBA This is a *warping-based algorithm*. Node points were placed at the centers of 16×16 blocks in the current frame. This formed a regular fixed mesh with square patches. In order for the mesh to cover the whole frame area, node points were also placed on the borders.

Backward node tracking was used to map the current node points to their matches in the reference frame. A continuous method was used to produce a continuous motion field. To ensure that the number of transmitted motion vectors is the same as that of the BMA, no motion vectors were transmitted for the border node points. Instead, each border node was assigned the motion vector of the closest inner node. However, to ensure that the borders of the current frame were mapped to the borders of the reference frame, border nodes at the corners of the frame were assigned zero motion vectors, the vertical component of a top or a bottom border nodal vector was set to zero, and the horizontal component of a left or a right border nodal vector was set to zero. The mesh geometry and nodal motion vectors are illustrated in Figure 5.3.

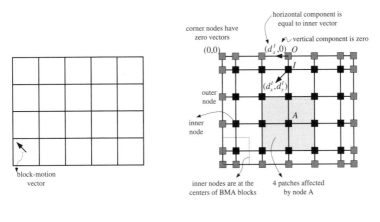

(a) BMA blocks   (b) WBA patches

Figure 5.3: BMA blocks and WBA patches

At the start of the node-tracking algorithm, the BMA described earlier was used to provide initial estimates of the inner nodal motion vectors. Those initial estimates were then refined using the iterative procedure of Sullivan and Baker [108]. In each iteration of this procedure, the nodes are processed sequentially, where the motion vector of a node is refined using a local search around the motion vector from the previous iteration while holding constant the motion vectors of its surrounding eight nodes. During this local search, the quality of a candidate motion vector is measured by calculating the distortion measure between all four patches connected to the node and their warped predictions from the reference frame. The local search is applied to a node only if its motion vector, or the motion vector of at least one of its surrounding nodes, was changed in the previous iteration. The local search used here examines the eight nearest candidate displacements centered around the displacement from the previous iteration. For each frame, 10 iterations were used to refine the nodal motion vectors.

During motion estimation and compensation, the bilinear spatial transformation is employed. This is implemented in the CGI [108] form (described in Section 5.2.7), where the motion vector used to compensate a pel within a patch is bilinearly interpolated, Equation (5.8), from the four nodal motion vectors at the vertices of the patch.

In BMA-HO and WBA algorithms, bilinear interpolation was used to obtain intensity values at subpel locations of the reference frame. In each algorithm, motion was estimated and compensated using original reference frames. Motion vectors were coded using the median predictor and the VLC table of the H.263 standard. The DFD signal was also transform encoded according to the H.263 standard and a quantization parameter of $QP = 10$. All quoted results refer to the luma components of sequences.

Table 5.1 compares the objective prediction quality of the preceding three algorithms when applied to the three test sequences with a frame skip of 3. The WBA outperforms the basic BMA by about 0.16–1.57 dB, depending on the sequence. However, the WBA fails to outperform the advanced BMA-HO

Table 5.1: Comparison between BMA and WBA in terms of objective prediction quality

	Average PSNR (dB) with a frame skip of 3		
	AKIYO	FOREMAN	TABLE TENNIS
BMA	39.88	27.81	29.06
WBA	41.45	29.09	29.22
BMA-HO	41.77	29.51	29.87

algorithm. In fact, the BMA-HO algorithm outperforms the WBA by about 0.32–0.65 dB.

Figure 5.4 compares the subjective prediction quality of the 45th frame of the 8.33-frames/s FOREMAN sequence when compensated using the preceding three algorithms. This figure shows that BMA-HO and WBA have approximately the same subjective quality and that both outperform the BMA. More importantly, this figure clearly shows the type of artefacts associated with each algorithm. The BMA suffers from the annoying blocking artefacts. Those artefacts are reduced by both the BMA-HO and the WBA algorithms. However, the BMA-HO algorithm has a low-pass filtering effect that smoothes sharp edges. This is due to the averaging (weighting) process during overlapped motion compensation. This effect is very clear at the edges of the helmet. The WBA, on the other hand, can suffer from warping artefacts. This is very clear at the top of the helmet, where part of the helmet was stretched to compensate

(a) Original 45th frame of FOREMAN at 8.33 f.p.s   (b) Compensated using BMA (28.06 dB)

(c) Compensated using BMA-HO (29.59 dB)   (d) Compensated using WBA (29.01 dB)

Figure 5.4: Comparison between BMA and WBA in terms of subjective prediction quality

uncovered background. In fact, poor compensation of covered and uncovered objects is one of the main disadvantages of the continuous warping-based method. In particular, the method performs poorly whenever there are objects disappearing from the scene because it can deform objects but cannot easily remove them completely [111].

Another obvious disadvantage of the continuous warping-based method is the lack of motion field segmentation. A number of methods have been proposed to overcome this problem. For example, Niewęgłowski and Haavisto [110] use adaptive motion field interpolation to introduce discontinuities within the nodal motion field. Adaptivity is achieved by switching between bilinear interpolation and nearest-neighbor interpolation of the nodal vectors at the vertices of a patch. The latter interpolation method effectively splits the motion field within the patch into four quadrants. A similar effect can be achieved by using a hierarchical (e.g., quad-tree) motion-based adaptive mesh [109, 120, 115, 123].

It is interesting at this point to compare the computational complexity of the preceding three algorithms. Table 5.2 compares the complexity of the three algorithms in terms of encoding time per frame. The results were obtained using the profiler of the Visual C++ 5.0 compiler run on a PC with Pentium 100-MHz processor, 64 MB of RAM, and a Windows 98 operating system. The results were averaged over 10 runs, where each run was used to encode the 8.33-frames/s FOREMAN sequence. Care should be taken when interpreting the results as they depend heavily on the implementation and the hardware platform.

The BMA requires about 2.16 seconds/frame. Most of this time (about 1.76 seconds) is consumed by the full-pel full-search block-matching motion estimation process.

The BMA-HO algorithm requires about 3.56 seconds/frame. This increase of about 1.4 seconds over the BMA is due mainly to two reasons. The half-pel refinement stage and the associated bilinear interpolation process increase the motion estimation time by about 0.98 seconds. In addition, the overlapping process increases the motion compensation time by about 0.42 seconds.

Table 5.2: Comparison between BMA and WBA in terms of computational complexity

	CPU time (in seconds) per frame when encoding FOREMAN at 8.33 f.p.s		
	BMA	BMA-HO	WBA
BMA motion estimation	1.76	2.74	1.86
WBA iterative refinement	0.00	0.00	116.00
Motion compensation	0.01	0.43	0.60
Others	0.39	0.38	0.37
Total	**2.16**	**3.56**	**118.83**

The WBA requires about 118.83 seconds/frame. This is a huge increase over both the BMA and the BMA-HO algorithms. This increase is due mainly to the iterative procedure used to refine the initial nodal vector estimates. Remember that in each iteration, for a single node to be refined, spatial transformation and bilinear interpolation have to be used to compensate the four patches connected to the node. There are a number of methods that can be used to alleviate this complexity. Examples are the use of fewer iterations per frame, the use of a line-scanning[1] technique to perform the spatial transformation, the use of a simpler interpolation method (e.g., nearest neighbor) or the use of a noniterative motion estimation algorithm, e.g. Ref. 130. Most of these methods, however, reduce the computational complexity at the expense of a reduced prediction quality.

5.4 Discussion

Block matching methods have always been criticized because of their simple uniform translational model. The argument against this model is that, in practice, a block can contain multiple moving objects and the motion is usually more complex than simple translation. The shortcomings of this model may appear as poor prediction quality for objects with nontranslational motion and also as blocking artefacts within motion-compensated frames. Warping-based methods employing higher-order motion models have been proposed in the literature as alternatives to block-matching methods. This chapter investigated the performance of warping-based methods and compared it to that of block-matching methods. The results of this comparison have shown that despite their improvements over basic block-matching methods, the use of warping-based methods in applications like mobile video communication may not be justifiable, due to the huge increase in computational complexity. In fact, similar (if not better) improvements can be obtained, at a fraction of the complexity, by simply augmenting basic block-matching methods with advanced techniques like subpel accuracy and overlapped motion compensation. One can argue that warping-based methods can also benefit from subpel accuracy and overlapped motion compensation, as shown in Refs. 113 and 117, but again this will further increase complexity. In addition to their high computational complexity, warping-based methods can suffer from warping artefacts,

---

[1]Once the motion vector of a pel (x, y) within a patch is interpolated from the four nodal vectors at the vertices of the patch, it can be shown that the motion vectors of the next pel in the line $(x + 1, y)$ and the next pel in the column $(x, y + 1)$ can be obtained by adding a simple update term. This is known as line scanning [107].

poor compensation of covered/uncovered background, and lack of motion field segmentation. Reducing the complexity of warping-based methods and including them in a hybrid WBA/BMA video codec are two possible areas of further research.

Chapter 6

Multiple-Reference Motion Estimation Techniques

6.1 Overview

To achieve high coding efficiency, Chapter 5 investigated an advanced motion estimation technique called warping-based motion estimation. This chapter considers another advanced technique, called *multiple-reference motion estimation*.

In *multiple-reference motion-compensated prediction* (MR-MCP), motion estimation and compensation are extended to utilize *more than one* reference frame. The reference frames are assembled in a *multiframe memory* (or buffer) that is maintained simultaneously at encoder and decoder. In this case, in addition to the spatial displacements, a motion vector is extended to also include a temporal displacement.

This chapter investigates the prediction gains achieved by MR-MCP. Particular emphasis is given to coding efficiency at very low bit rates. More precisely, the chapter attempts to answer the following question: *Is the use of additional bit rate to transmit the extra temporal displacement justifiable in terms of an improved rate-distortion performance?* The chapter also examines the properties of the multiple-reference block-motion field and compares them to those of the single-reference case.

The rest of the chapter is organized as follows. Section 6.2 briefly reviews multiple-reference motion estimation techniques. Section 6.3 concentrates on the long-term memory multiple-reference motion estimation technique. The section starts by examining the properties of multiple-reference block-motion fields and compares them to those of single-reference fields. It then investigates the prediction gains and the efficiency of the long-term memory technique at very low bit rates. The chapter concludes with a discussion in Section 6.4.

6.2   Multiple-Reference Motion Estimation: A Review

In *multiple-reference motion-compensated prediction* (MR-MCP), motion estimation and compensation are extended to utilize *more than one* reference frame. The reference frames are assembled in a *multi-frame memory* (or buffer) that is maintained simultaneously at encoder and decoder. In this case, in addition to the spatial displacements (d_x, d_y), a motion vector is extended to also include a *temporal displacement* d_t. This is the *index* into the multiframe memory. The process of MR-MCP is illustrated in Figure 6.1.

The main aim of MR-MCP is to improve coding efficiency. Thus, the reference generation block in Figure 6.1(a) can utilize any technique that provides useful data for motion-compensated prediction. Examples of such techniques are reviewed in what follows.

A number of MR-MCP techniques have been proposed for inclusion within MPEG-4. Examples are *global motion compensation* (GMC) [131, 132], *dynamic sprites* (DS) [132], and *short-term frame memory/long-term frame memory* (STFM/LTFM) prediction [133]. In these techniques, MCP is performed using two reference frames. The first reference frame is always the past decoded frame, whereas the second reference frame is generated using different methods. In GMC, the past decoded frame is warped to provide the second reference frame. The technique of DS is a more general case of GMC. In DS, past decoded frames are warped and blended into a sprite memory. This sprite memory is used to provide the second reference frame. In STFM/LTFM two frame memories are used. The STFM is used to store the past decoded frame, whereas the LTFM is used to store an earlier decoded frame. The LTFM is updated using a refresh rule based on scene-change detection. Both DS and STFM/LTFM can benefit from another MR-MCP technique, which is background memory prediction [134].

Similar to the STFM/LTFM is the reference picture selection (RPS) mode included in annex N of H.263+ (refer to Chapter 3). In this mode, switching to a different reference picture can be signaled at the picture level. It should be pointed out, however, that this option was designed for error resilience rather than for coding efficiency. Its main function is to stop error propagation due to transmission errors.

Probably the most significant contributions to the field of MR-MCP are those made by Wiegand and Girod *et al.* [135–141]. They noted [135, 136] that long-term statistical dependencies in video sequences are not exploited by existing video standards. Thus, they proposed to extend motion estimation and compensation to utilize several past decoded frames. They called this technique *long-term-memory motion-compensated prediction* (LTM-MCP). They demonstrated that the use of this technique can lead to significant improvements in coding efficiency.

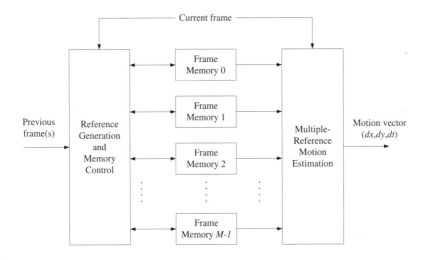

(a) Multiple-reference motion estimation

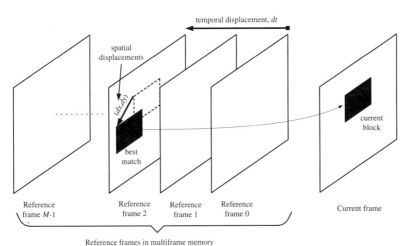

(b) Multiple-reference motion compensation

Figure 6.1: Multiple-reference motion-compensated prediction

In Ref. 137 they proposed to use multiple global motion models to generate the reference frames. Thus, reference frames in this case are warped versions of the previously decoded frame using polynomial motion models. This can be seen as an extension to GMC, where, in addition to the most dominant global motion, less dominant motion is also captured by additional motion parameter sets. In order to determine the multiple models, a robust clustering method based on the iterative application of the least median of squares estimator is employed. This model estimation method is computationally expensive. In Ref. 138 they proposed an alternative method in which the past decoded frame is split into blocks of fixed size. Each block is then used to estimate one model using translational block matching followed by a gradient-based affine refinement. In addition to reduced complexity, this method leads to higher prediction gains.

In Ref. 139 they have demonstrated that combining the LTM-MCP method of Refs. 135 and 136 with the multiple GMC method of Ref. 138 can lead to further coding gains.

Recently, MR-MCP has been included in the enhanced reference picture selection (ERPS) mode (annex U) of H.263++ (refer to Chapter 3).

6.3 Long-Term Memory Motion-Compensated Prediction

As already discussed, there are many MR-MCP techniques. The main difference between those techniques is in the way they generate the reference frames. The simplest and least computationally complex approach is the LTM-MCP technique, where past decoded frames are assembled in the multiframe memory. This chapter will therefore concentrate on the LTM-MCP technique. More complex techniques, such as multiple GMC, may not be suitable for computationally constrained applications such as mobile video communication.

There are many ways to control the multiframe memory in the LTM-MCP technique. The simplest approach is to use a *sliding-window* control method. Assuming that there are M frame memories: $0\ldots M-1$, then the most recently decoded past frame is stored in frame memory 0, the frame that was decoded M time instants before is stored in frame memory $M-1$, and so on. In the next time instant, the window is moved such that the oldest frame is dropped from memory, the contents of frame memories $0\ldots M-2$ are shifted to frame memories $1\ldots M-1$, and the new past decoded frame is stored in frame memory 0. According to this arrangement the new motion vector component is in the range $0 \leq d_t \leq M-1$, where $d_t = 0$ refers to the most recent reference frame in memory. This sliding-window technique will be adopted throughout this chapter.

6.3.1 Properties of Long-Term Block-Motion Fields

This subsection investigates the properties of long-term block-motion fields and compares them to those of single-reference block-motion fields. All illustrations in this subsection were generated using a full-pel full-search long-term memory block-matching algorithm applied to the luma component of the FOREMAN sequence with blocks of 16×16 pels, a maximum allowed displacement of ± 15 pels, SAD as the distortion measure, restricted motion vectors, and original reference frames.

Property 6.3.1.1 *The distribution of the long-term memory spatial displacements (d_x, d_y) is center-biased.* This is evident from Figure 6.2, which shows the distribution of the relative frequency of occurrence of the spatial displacements d_x (Figure 6.2(a)) and d_y (Figure 6.2(b)). Note that this is similar to the single-reference case ($M = 1, \text{skip} = 1$), although in the case of multiple-reference ($M = 50, \text{skip} = 1$), the distribution is slightly more spread, which indicates that longer displacements are slightly more probable. This distribution is even more spread at higher frame skips, ($M = 50, \text{skip} = 4$).

Property 6.3.1.2 *The distribution of the long-term memory temporal displacement d_t is zero-biased.* This is evident from Figure 6.3, where the temporal displacement $d_t = 0$ (which refers to the most recent reference frame

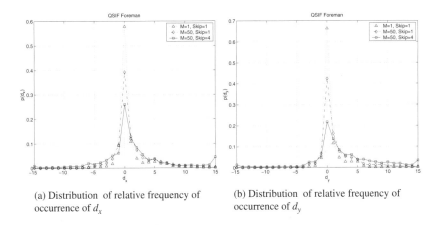

(a) Distribution of relative frequency of occurrence of d_x

(b) Distribution of relative frequency of occurrence of d_y

Figure 6.2: Center-biased distribution of the long-term memory spatial displacements (d_x, d_y)

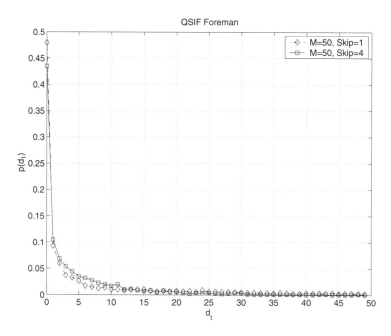

Figure 6.3: Zero-biased distribution of the long-term memory temporal displacement d_t

in memory) has the highest frequency of occurrence; and as the temporal displacement increases, its frequency of occurrence decreases. Note that this distribution becomes more spread at higher frame skips, which indicates that the selection of older reference frames becomes slightly more probable.

Property 6.3.1.3 *The long-term memory block-motion field is smooth and varies slowly.* In other words, there is high correlation between the motion vectors of adjacent blocks. This is evident from Figure 6.4, which shows the distribution of the difference between the current vector **C** and its left neighbor **L**. This is shown for the three components: d_x (Figure 6.4(a)), d_y (Figure 6.4(b)), and d_t (Figure 6.4(c)). All three distributions are biased toward a zero difference, which indicates high correlation. Note that this correlation is slightly less in the multiple-reference case ($M = 50, \text{skip} = 1$), compared to the single-reference case ($M = 1, \text{skip} = 1$). This correlation is further reduced at higher frame skips, ($M = 50, \text{skip} = 4$).

In general, it can be concluded that moving from a single-reference system to a multiple-reference system does not significantly change the properties of the block-motion field.

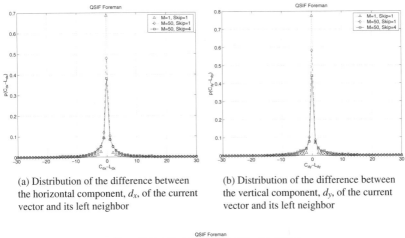

(a) Distribution of the difference between the horizontal component, d_x, of the current vector and its left neighbor

(b) Distribution of the difference between the vertical component, d_y, of the current vector and its left neighbor

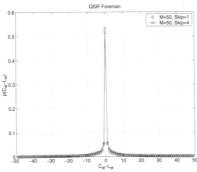

(c) Distribution of the difference between the temporal component, d_t, of the current vector and its left neighbor

Figure 6.4: Highly correlated long-term memory block-motion field

6.3.2 Prediction Gain

This subsection evaluates the prediction gain achieved by LTM-MCP. All results were generated using full-pel full-search long-term memory block matching with blocks of 16×16 pels, a maximum allowed displacement of ± 15 pels, SAD as the distortion measure, restricted motion vectors, and original reference frames. All quoted results refer to the luma components of sequences.

Figure 6.5 shows the performance of LTM-MCP when applied to the three QSIF sequences AKIYO, FOREMAN, and TABLE TENNIS with different memory sizes and different frame skips. It is immediately evident from this figure that significant prediction gains are achieved when utilizing more than one

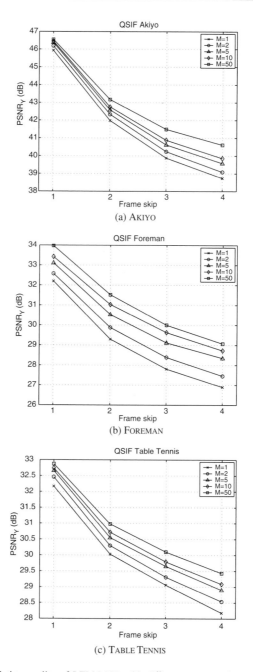

(a) AKIYO

(b) FOREMAN

(c) TABLE TENNIS

Figure 6.5: Prediction quality of LTM-MCP with different memory sizes and frame skips

reference frame. For example, at a frame skip of 4, the prediction gain when using a multiframe memory of size $M = 50$ frames is 1.87 dB for AKIYO, 2.17 dB for FOREMAN, and 1.25 dB for TABLE TENNIS, compared to single-reference prediction (i.e., $M = 1$). Such prediction gains are mainly due to the long-term statistical dependencies of video sequences. Examples of such dependencies are the repetitions of sequence content due to uncovered objects or objects reappearing in the sequence. An interesting point to note here is that the prediction gains increase with increased frame skip. For example, for AKIYO when going from $M = 1$ to $M = 50$, the prediction gain is 0.62 dB at a frame skip of 1 and 1.87 dB at a frame skip of 4. This may be due to the fact that as the frame skip increases, successive frames get more decorrelated. This increases the chance that a frame other than the immediately preceding one will be chosen and, consequently, gives more chance to benefit from long-term memory prediction. In Ref. 136, the benefits of extending LTM-MCP to half-pel accuracy are discussed. It is shown that further prediction gains can be achieved by moving from full- to half-pel accuracy. This "accuracy gain" is comparable to that in the case of single-reference prediction.

It should be emphasized that the improved prediction quality of LTM-MCP is achieved at the expense of:

1. Increased memory requirements at both the encoder and the decoder.

2. Additional bit rate to transmit the new extra components, d_t, of motion vectors.

3. Increased computational complexity at the encoder.

Item 1 is not a major drawback due to the rapid drop in the price of memory chips, item 2 will be investigated further in Section 6.3.3, whereas a possible solution for item 3 will be proposed in Chapter 8.

6.3.3 Efficiency at Very Low Bit Rates

As already discussed in Section 6.1, LTM-MCP extends the motion vector of a block by a third component, d_t. This is the temporal displacement or the index into the multiframe memory. Obviously, the transmission of this extra component incurs an additional bit rate compared to the single-reference case. This additional bit rate has to be justified in terms of an improvement in the rate-distortion (R-D) performance. This subsection investigates the R-D performance of the LTM-MCP technique. Particular emphasis is given to the efficiency of this technique at the very low bit rates typical of mobile video communication. Four H.263-like encoders were implemented:

SR This is a single-reference encoder. It uses full-pel full-search block matching with macroblocks of 16×16 pels, a maximum allowed spatial displacement of ± 15 pels, SAD as the distortion measure, restricted motion vectors, and reconstructed reference frames. Motion vectors are coded using the median predictor and the VLC table of the H.263 standard. The frame signal (in case of INTRA) and the DFD signal (in case of INTER) are transform encoded according to the H.263 standard. The encoder does not employ rate-constrained motion estimation and mode decision. Thus, motion estimation simply chooses the motion vector that minimizes the SAD measure without any bit-rate considerations. The INTRA/INTER decision is based on heuristic thresholds and is given by the following [142]:

$$\text{INTRA mode is chosen if} \quad A < (\text{SAD}(\mathbf{d}) - 500), \quad (6.1)$$

where

$$A = \sum_{(x,y) \in \mathscr{B}} |f_c(x, y) - \bar{B}| \quad (6.2)$$

and

$$\bar{B} = \frac{\sum_{(x,y) \in \mathscr{B}} f_c(x, y)}{256}, \quad (6.3)$$

where $\mathbf{d} = (d_x, d_y)$ is the motion vector of macroblock $\mathscr{B}$ in the current frame f_c and $\text{SAD}(\mathbf{d})$ is the SAD between the macroblock in the current frame and a corresponding macroblock in the reference frame shifted by $\mathbf{d}$.

SR-RC This is a single-reference rate-constrained encoder. It is the same as SR, but it uses rate-constrained motion estimation and mode decision as defined in the high-complexity mode of the H.263 test model, near-term, version 10 (TMN10) [142]. In this mode, motion estimation chooses the motion vector that minimizes the following Langrangian cost function:

$$J_{\text{MOTION}} = D_{\text{MOTION}} + \lambda_{\text{MOTION}} R_{\text{MOTION}}, \quad (6.4)$$

where D_{MOTION} is the SAD between the macroblock in the current frame and the corresponding macroblock in the reference frame shifted by $\mathbf{d}$, R_{MOTION} is the number of bits used to encode the motion vector $\mathbf{d}$, and λ_{MOTION} is a Lagrange multiplier related to the quantization parameter QP using

$$\lambda_{\text{MOTION}} = 0.92 \times \text{QP}. \quad (6.5)$$

To decide the mode, two Langrangian cost functions, one for each mode, are calculated as follows:

$$J_{\text{INTRA}} = D_{\text{INTRA}} + \lambda_{\text{MODE}} R_{\text{INTRA}}, \quad (6.6)$$

$$J_{\text{INTER}} = D_{\text{INTER}} + \lambda_{\text{MODE}} R_{\text{INTER}}, \quad (6.7)$$

where D_{INTRA} is the SSD between the current macroblock and its INTRA encoded reconstruction and R_{INTRA} is the number of bits used to INTRA encode the current macroblock. Similar definitions also apply for D_{INTER} and R_{INTER}, but they are calculated by INTER encoding the current macroblock. In both equations, λ_{MODE} is a Lagrange multiplier related to the quantization parameter QP using

$$\lambda_{\text{MODE}} = 0.85 \times \text{QP}^2. \tag{6.8}$$

The mode with the minimum cost function is chosen as the mode of the current macroblock. Note that, in this case, a macroblock needs to be encoded twice before being able to decide its mode. This increases the complexity of the encoder. A more detailed description of this rate-constrained motion estimation and mode decision method can be found in Ref. 143.

MR This is a multiple-reference encoder with no rate constraints. Thus, it is the same as SR, but it uses long-term memory motion-compensated prediction.

MR-RC This is a multiple-reference rate-constrained encoder. Thus, it is the same as SR-RC, but it uses long-term memory motion-compensated prediction.

The preceding encoders were tested using the three QSIF test sequences AKIYO, FOREMAN, and TABLE TENNIS. The frame skip parameter was set to 3 to achieve low bit rates. To generate a range of bit rates, the quantization parameter QP was varied over the range 5–30 in steps of 5. This means that each encoder was used to encode a given sequence six times. Each time, QP was held constant over the whole sequence (i.e., no rate control was used). The first frame was always INTRA encoded. The INTRA bits of the first frame were included in the bit-rate calculations, and no header bits were generated. All quoted results refer to the luma components of sequences. For MR and MR-RC, sliding-window control was used to maintain a long-term memory of size $M = 50$ frames. The VLC codewords in Table 6.1 were used to encode[1] the temporal components d_t of the long-term motion vectors.

Figures 6.6, 6.7, and 6.8 show the R-D performance of the preceding encoders for the three test sequences. Note that both single-reference and

---

[1]For example, since $d_t = 4$ is in the range (3:6), then according to Table 6.1 it will be encoded using a 5-bit codeword. This codeword is derived as follows. With reference to the start of its range, $d_t = 4$ is represented by $d_t - 3 = 4 - 3 = 1$. Thus, $x_1 x_0 = 01$ and the codeword is given by $0x_1 1x_0 0 = 00110$.

Table 6.1: VLC codewords for encoding the temporal displacement d_t. Reproduced from Ref. 140

d_t	Bits	Codeword
0	1	1
"x_0" $+ 1$ (1:2)	3	$0x_00$
"x_1x_0" $+ 3$ (3:6)	5	$0x_11x_00$
"$x_2x_1x_0$" $+ 7$ (7:14)	7	$0x_21x_11x_00$
"$x_3x_2x_1x_0$" $+ 15$ (15:30)	9	$0x_31x_21x_11x_00$
"$x_4x_3x_2x_1x_0$" $+ 31$ (31:62)	11	$0x_41x_31x_21x_11x_00$

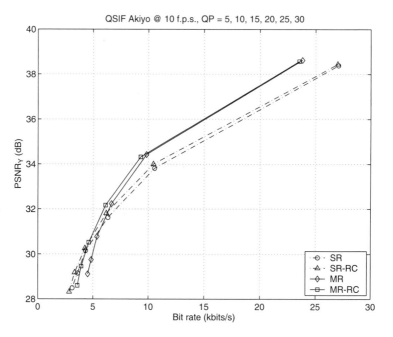

Figure 6.6: R-D performance of different single- and multiple-reference (with $M = 50$) encoders when encoding QSIF AKIYO at 10 frames/s

multiple-reference encoders benefit from the use of rate-constrained motion estimation and mode decision. Those benefits are more evident in high-movement sequences, where the use of more bits to encode the longer motion vectors has to be justified and controlled. It should be pointed out, however, that such benefits are achieved at the expense of increased computational complexity.

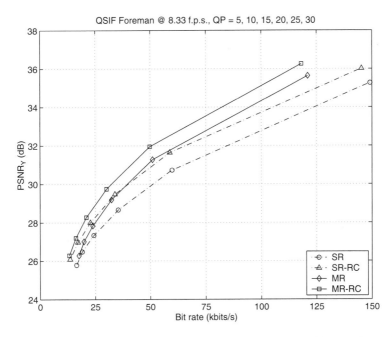

Figure 6.7: R-D performance of different single- and multiple-reference (with $M = 50$) encoders when encoding QSIF FOREMAN at 8.33 frames/s

Due to the additional bit rate generated by the temporal components d_t, the use of rate-constrained motion estimation and mode decision is essential in the case of multiple-reference encoders. A single-reference rate-constrained encoder (SR-RC) can outperform a multiple-reference encoder with no rate constraints (MR). This is evident at very low bit rates in Figures 6.6 and 6.7 and at all bit rates in Figure 6.8. In fact, at very low bit rates, even a single-reference encoder with no rate constraints (SR) can sometimes outperform the multiple-reference encoder (MR).

The best overall performance is achieved by the multiple-reference rate-constrained encoder (MR-RC). The benefits of this encoder become more evident as the bit rate increases. Note, however, that this improved performance is at the expense of a significant increase in computational complexity. This increase is due to the use of more than one reference frame during motion estimation and also to the use of rate-constrained motion estimation and mode decision. Note, also, that at extremely low bit rates a similar performance can be achieved by the less complex (SR-RC) encoder. Thus, at such bit rates the use of LTM-MCP is not justifiable.

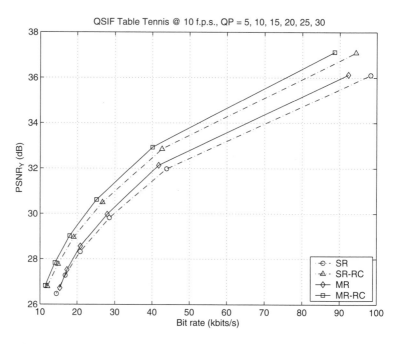

Figure 6.8: R-D performance of different single- and multiple-reference (with $M = 50$) encoders when encoding QSIF TABLE TENNIS at 10 frames/s

6.4 Discussion

Higher coding efficiency is one of the main requirements for mobile video communication. One way to achieve higher coding efficiency is to use advanced motion estimation techniques. One of the promising advanced techniques is multiple-reference motion-compensated prediction (MR-MCP).

This chapter reviewed the main efforts in the field of MR-MCP. It then investigated the performance of the long-term memory motion-compensated prediction (LTM-MCP) technique. It was found that this technique provides significant prediction gains compared to the single-reference case. It was realized, however, that such prediction gains are achieved at the expense of an additional bit rate to transmit one extra temporal component per motion vector. This additional bit rate has to be justified in terms of an improved rate-distortion (R-D) performance. An investigation into the R-D performance of LTM-MCP codecs revealed that the use of rate-constrained motion estimation and mode decision is important for the success of such techniques. Without rate constraints, the R-D performance of the LTM-MCP technique can, at very low bit rates, drop below that of single-reference codecs. Combined with rate constraints, the LTM-MCP technique provides a superior R-D performance, which becomes more evident as the bit rate increases.

The chapter investigated the properties of long-term memory block-motion fields. It was found that the distribution of the long-term memory spatial displacements is center-biased. This distribution becomes more spread with increased frame memory size and frame skip. It was also found that the distribution of the long-term memory temporal displacement is zero-biased. Again, this distribution becomes more spread with increased frame memory size and frame skip. The investigation revealed also that the long-term memory block-motion field is highly correlated. In general, it was concluded that moving from a single-reference system to a multiple-reference system does not significantly change the properties of the block-motion field.

Part III
Computational Complexity

In mobile terminals, processing power and battery life are very limited and scarce resources. Given the significant amount of computational power required to process video, the use of reduced-complexity techniques is essential.

Motion estimation is the most computationally intensive process in a typical video codec. In fact, the computational complexity of this process is greater than that of all the remaining encoding steps combined. Thus, by reducing the complexity of this process, the overall complexity of the codec can be reduced.

This part contains two chapters. Chapter 7 reviews reduced-complexity motion estimation techniques. The chapter uses implementation examples and profiling results to highlight the need for reduced-complexity motion estimation. It then reviews some of the main reduced-complexity block-matching motion estimation techniques. The chapter then presents the results of a study comparing the different techniques.

Chapter 8 gives an example of the development of a novel reduced-complexity motion estimation technique. The technique is called the *simplex minimization search* (SMS). The development process is described in detail, and the SMS technique is then tested within an isolated test environment, a block-based H.263-like codec, and an object-based MPEG-4 codec. In an attempt to reduce the complexity of multiple-reference motion estimation (investigated in Chapter 6), the chapter then extends the SMS technique to the multiple-reference case. The chapter presents three different extensions (or algorithms) representing different degrees of compromise between prediction quality and computational complexity.

Chapter 7

Reduced-Complexity Motion Estimation Techniques

7.1 Overview

As already discussed, one of the main requirements for mobile video communication is *reduced-complexity*. It is not difficult to show that the high computational complexity of a typical video codec is due mainly to the *motion estimation* process. Thus, by reducing the complexity of this process, the overall complexity of the codec can be reduced. This chapter reviews reduced-complexity motion estimation techniques. In particular, the chapter concentrates on reduced-complexity block-matching motion estimation (BMME) techniques. The chapter also presents the results of a study comparing different reduced-complexity BMME techniques.

The rest of the chapter is organized as follows. Section 7.2 uses implementation examples and profiling results to highlight the need for reduced-complexity motion estimation. Sections 7.3–7.7 review the main categories of reduced-complexity BMME algorithms. Section 7.8 presents the results of a study comparing the different categories. The chapter concludes with a discussion in Section 7.9.

7.2 The Need for Reduced-Complexity Motion Estimation

Processing digital video requires a significant amount of computational power. This represents one of the main challenges for real-time mobile video communication, where processing power and battery life are scarce resources. For example, an MPEG-4 simple profile codec has recently been implemented on Texas Instruments' TMS320C541 40-MHz processor [5].[1] Profiling results show that this codec cannot achieve real-time processing even when using SQCIF sequences. It can encode only about 1 frame/s, and it can decode only about 20 frames/s. Another example is the implementation of the H.263 baseline mode on the more powerful TMS320C62 200-MHz processor, as described in Ref. 6. Again, this implementation cannot achieve real-time processing, for it only can encode about 5 QCIF frames/s.

Looking at the building blocks of a typical video codec, it is not difficult to realize that this huge computational complexity is due mainly to the motion estimation process. As already discussed, most video codecs estimate motion using the *block-matching motion estimation* (BMME) algorithm. The most straightforward BMME algorithm is the *full-search* (FS) algorithm, sometimes referred to as the *exhaustive search* or the *brute-force search*. This algorithm is guaranteed to find the best-match block because it exhaustively searches over all possible blocks (search locations or candidate motion vectors) within the search window. The algorithm produces the best possible prediction quality. This is, however, at the expense of a huge computational complexity. For example, for a CIF video sequence encoded at 30 frames/s with 16×16 blocks, maximum displacement of ± 15 pels, and SAD as the distortion measure, a direct implementation of a full-pel FS-BMME algorithm requires about 6×10^9 integer additions and subtractions, 3×10^9 magnitude operations, and 11×10^6 comparisons per second. In fact, the computational complexity of this motion estimation process is greater than that of all the remaining encoding steps combined. This is clear from Table 7.1, which shows profiling results[2] of the baseline mode of Telenor's H.263 encoder [144] when used to encode the QCIF FOREMAN sequence at 64 kbits/s. In this case, the motion estimation process[3] consumes about 70% of the overall encoding time.

Because of this high computational complexity, motion estimation has become a bottleneck problem in many applications, e.g., mobile video terminals and software-based video codecs, especially if real-time video coding is required. This has motivated the development of a number of fast motion estimation algorithms since the early 1980s. In fact, recent advances in video coding not only highlight the importance of such algorithms, but even call for further research into the area of reduced-complexity motion estimation. For example, HDTV and multiple-reference motion estimation (discussed

---

[1] According to Ref. 5, about half of all mobile phones currently use a 'C54x processor.

[2] The results were obtained using the profiler of the Visual C++ 5.0 compiler run on a PC with a Pentium 100-MHz processor, 64 MB of RAM, and a Windows 98 operating system.

[3] The baseline mode of Telenor's H.263 codec uses block matching with 16×16 blocks, SAD as the distortion measure, and ± 15 pels maximum displacement. Full-pel accuracy is first obtained using full search. This is then refined to half-pel accuracy.

Table 7.1: Profiling results of Telenor's H.263 baseline mode when used to encode QCIF FOREMAN at 64 kbits/s

Function	CPU Time (ms)	Percentage (%)
Motion estimation	240,354	69.7
Input/output	32,552	9.4
DCT/IDCT	29,412	8.5
Others	42,353	12.4
Total	344,671	100.0

in Chapter 6) have a computational complexity that is several orders of magnitude higher than that shown in the preceding examples. The former uses higher spatial resolutions and larger search windows, and the latter extends the search over several reference frames.

The following sections of this chapter review the main categories of reduced-complexity BMME algorithms. Although each category can be used on its own, careful encoder design can utilize different combinations to achieve higher speed-up ratios.

7.3 Techniques Based on a Reduced Set of Motion Vector Candidates

Instead of searching over all possible blocks within the search window, this category restricts the search over a selected subset of the blocks. Most algorithms in this category are, implicitly or explicitly, based on the *unimodal error surface assumption* [54], which states that *the block distortion measure (BDM) increases monotonically as the search location moves away from the best-match location.* Therefore, the search starts by evaluating the BDM at locations coarsely spread over the search window according to some predefined uniform pattern. This is then repeated with finer resolution (i.e., smaller spread) around the search location with the minimum BDM from the preceding step.

The first algorithm reported in this category was the *two-dimensional logarithmic* (TDL) search proposed in 1981 by Jain and Jain [54]. Figure 7.1 shows an example of the TDL search with a maximum displacement of $d_m = 6$ pels. The search is initialized at the origin of the search window with a suitable step size (i.e., the spacing between the search locations). In each step of the TDL search, the BDM is evaluated at five search locations. In the given example, the search locations $(0,0)$, $(+2,0)$, $(-2,0)$, $(0,+2)$, $(0,-2)$ form

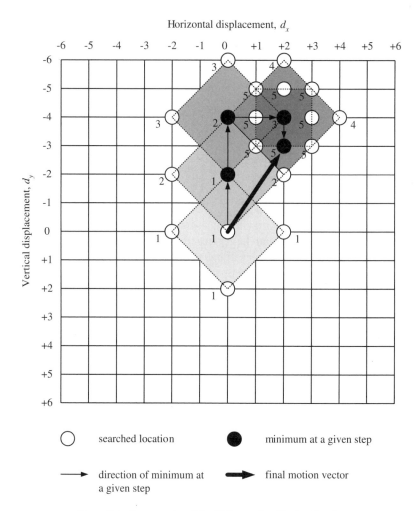

Figure 7.1: An example of the TDL search with $d_m = 6$ pels

the search pattern of the first step. At each step, the search pattern is centered around the minimum of the previous step. In the given example, the minimum in the first step is at $(0, -2)$. Thus, the search pattern in the second step is centered around this minimum location. The step size is reduced by a factor of 2 if the minimum is in the center of the search pattern or at the boundary of the search window. In the fourth step of the given example, the minimum is at $(+2, -4)$, which is the center of the search pattern. Therefore, the spacing between the search locations is halved in the fifth step. Since halving the distance between the search locations gives a step size of 1, this indicates that this is the final step in the search. In this final step, BDM is evaluated at the minimum from the previous step and also at its eight nearest neighbors. In the given example, the final motion vector is $(+2, -3)$.

Since the introduction of the TDL search, a large number of similar algorithms have been proposed. Examples are the *three-step search* (TSS) [145], the *one-at-a-time search* (OTS) [146], the *conjugate-directions search* (CDS) [146], the *cross-search algorithm* (CSA) [147], the *genetic motion search* (GMS) [148], and the *diamond search* (DS) [149–151], to mention a few. The appendix gives a detailed description of some of these algorithms.

Compared to other techniques, this category of techniques provides a relatively high speed-up ratio and has, therefore, received most of the attention. However, as is shown later, the unimodal error surface assumption does not always hold true, and such algorithms can easily be trapped in local minima, resulting in a suboptimal prediction quality.

7.4 Techniques Based on a Reduced-Complexity Block Distortion Measure

In this category, reduced-complexity is achieved by employing a reduced-complexity BDM. As already discussed in Chapter 4 (Section 4.6.1), most implementations prefer the SAD measure, due to its reduced-complexity and good prediction quality. A number of other reduced-complexity BDMs have also been proposed in the literature. Examples are the *pel difference classification* (PDC) [152], the *minimized maximum* (MiniMax) error [153], the *reduced-bits mean absolute difference* (RBMAD) [154], *integral projections* [155], and *one-bit/pixel* [156], to mention a few. Most of these measures were designed specifically for efficient hardware and VLSI implementation, but their prediction quality is not as good as the SSD or the SAD measures.

Another type of algorithms in this category reduces the complexity of the BDM by subsampling the matched blocks. Obviously, since only a fraction of the pels is used in the matching process, this category does not guarantee to find the best match, even when combined with full search. Koga *et al.* [145] subsample the matched blocks by a factor of 2 both horizontally and vertically (i.e. 4:1 subsampling), reducing the complexity of the BDM by a factor of 4.

Instead of using a uniform subsampling pattern, Liu and Zaccarin [157] use alternating subsampling patterns. The patterns are alternated over the searched locations in such a way that effectively all pels of a block contribute to the matching process. This method is illustrated in Figure 7.2. Figure 7.2(a) shows an 8×8 block of pels. Four 4:1 subsampling patterns are defined in this block. For example, subsampling pattern **A** consists of all pels labeled **a** in the block.

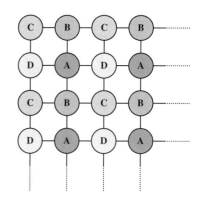

(a) Four 4:1 subsampling patterns

(b) Alternating schedule of four subsampling patterns over the search window

Figure 7.2: Reduced-complexity BDM using the alternating subsampling patterns of Liu and Zaccarin [157]

Similarly, patterns **B**, **C**, and **D** consist of all the **b**, **c**, and **d** pels, respectively. Figure 7.2(b) shows part of the search window in the reference frame. Each circle in this figure represents a search location (i.e., a candidate block) in the window. During motion estimation, search locations labeled **A** use the subsampling pattern **A**, and so on. For each of the four subsampling patterns, the motion vector with the minimum BDM over the locations where that pattern is used is selected. For each of the four selected motion vectors, the BDM is evaluated, but this time without subsampling. The vector that achieves the minimum BDM is selected as the motion vector of the block. Compared to the approach of Koga *et al.*, this approach achieves approximately the same reduction in complexity, but with better prediction quality.

Chan and Siu [158] vary the number of pels in the subsampling pattern according to block details. Thus, fewer pels are used for uniform blocks and more pels are used for high-activity blocks. In this algorithm, the reduction in complexity varies between blocks and the prediction quality is generally better than that of Liu and Zaccarin.

7.5 Techniques Based on a Subsampled Block-Motion Field

This category is based on the fact that block-motion fields of typical video sequences are usually smooth and vary slowly (as was shown in Section 4.6.7).

In other words, it is very common to find neighboring blocks with identical or nearly identical motion vectors. Thus, in this category, a subsampled block-motion field is first obtained by estimating the motion vectors for a fraction of the blocks in the frame. This field is then appropriately interpolated to determine the motion vectors of the remaining blocks.

Liu and Zaccarin [157] use a checkerboard subsampling pattern and estimate the motion vectors for half of the blocks (i.e., a 2:1 subsampled motion field) using full search. Then they estimate the motion vectors of the other half using a limited search that examines only four candidate motion vectors. Those candidates are the four surrounding motion vectors that were estimated using full search. For example, in Figure 7.3(a) the motion vectors of blocks **B**, **C**, **D**, and **E** are estimated using full search. Only those four vectors are then used as candidates when estimating the motion of block **A**. This algorithm reduces complexity by roughly a factor of 2, with only a slight loss in prediction quality.

Another algorithm was also proposed by Liu and Zaccarin in Ref. 157. In this algorithm, each block is divided into four subblocks. Motion is first estimated, using full search, for one subblock in each block, say, the top-left subblock. The motion vectors of the remaining subblocks are then estimated using a limited search with candidates from the neighboring full-search motion vectors. For example, in Figure 7.3(b) the motion vectors of subblocks **A**, **B**, **C**, and **D** are estimated using full search. Only those four vectors are then used as candidates when estimating the motion vectors of subblocks **a**, **b**,

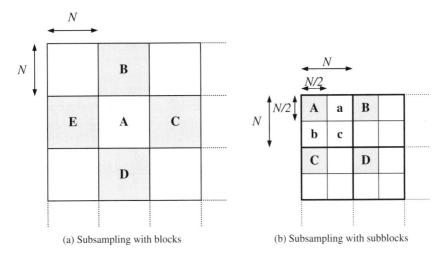

(a) Subsampling with blocks   (b) Subsampling with subblocks

Figure 7.3: Reduced-complexity using the subsampled motion fields of Liu and Zaccarin [157]

and **c**. This algorithm reduces complexity by roughly a factor of 4. Since smaller blocks are employed, the algorithm provides better prediction quality than full-search with original size blocks. This is, however, at the expense of a larger motion overhead.

7.6   Hierarchical Search Techniques

This category uses a multiresolution representation of video. The basic idea is to perform motion estimation at each level successively, starting with the lowest resolution level. Thus, motion estimation is first performed at the lowest resolution level to obtain a rough estimate of the motion vector. This vector is then passed to the next-higher resolution level to serve as an initial estimate. Motion estimation at the higher resolution level is then used to refine this initial estimate. This process is repeated until the highest resolution level is reached. At lower resolution levels, smaller blocks are used for block matching. This reduces the complexity of calculating BDMs. At higher-resolution levels, smaller search ranges are used since motion estimation starts from a good initial estimate. This reduces the number of locations to be searched. Both factors (i.e., smaller blocks at low resolutions and smaller search ranges at high resolutions) contribute to reducing the overall complexity of the search. Note that when reducing the resolution of the searched frames, the motion speed is also reduced. This makes hierarchical techniques particularly useful for estimating, with reduced complexity, high motion content. Examples of hierarchical motion estimation algorithms are reported in Refs. 159 and 160.

Figure 7.4 shows an example of a three-level hierarchical motion estimation technique applied to a QCIF sequence. In this case, the current frame is first used to generate three current frames with the resolutions: 44×36, 88×72, and 176×144. Each resolution level is a low-pass filtered and subsampled version of the next-higher resolution level. The resulting representation is called a *mean pyramid*. The same process is also applied to the reference frame (i.e., the previous frame). Motion estimation starts at the lowest resolution level with a block size of 4×4 pels and a search range of ± 3 pels. The estimated motion vector of a block in this resolution is scaled up by a factor of 2 (i.e., the scaled vector will have a maximum range of $2 \times (\pm 3) = \pm 6$ pels) and then be passed to the corresponding block in the next-higher resolution level. Motion estimation in the next-higher resolution level uses a block size of 8×8 pels and a search range of ± 1 pel around the propagated vector from the lower resolution level. This produces a motion vector with a maximum range of $(\pm 6) + (\pm 1) = \pm 7$ pels, which is again scaled up by a factor of 2 (to a maximum range of ± 14) and propagated to the next-higher resolution level. In this level, a block size of 16×16 pels is used with a search range

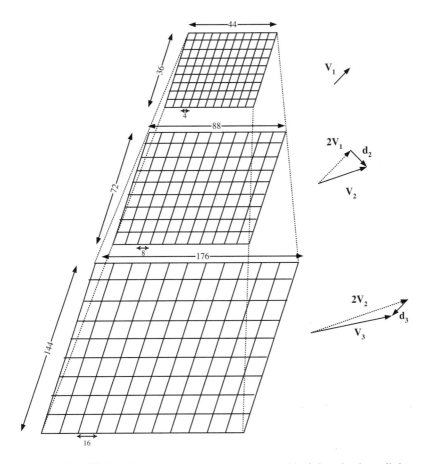

Figure 7.4: Hierarchical motion estimation using a mean pyramid of three levels applied to a QCIF frame

of ± 1 pel around the propagated vector from the lower resolution level. This gives a final vector with a maximum range of ± 15 pels.

There are many variants to hierarchical motion estimation. Some techniques use the same frame size in all levels of the hierarchy, with larger block sizes at lower levels. Other techniques use the same block size in all levels of the hierarchy, with subsampled frames at lower levels. In both cases, any level will have fewer blocks than the next higher level. Thus, a motion vector estimated at one level will be propagated to more than one block in the higher level.

In addition to reduced-complexity and robust estimation of high-motion content, hierarchical motion estimation algorithms are also reported to provide more homogeneous block-motion fields and a better representation of the true motion in the frame [159]. The latter property is particularly important for motion-compensated interpolation.

7.7 Fast Full-Search Techniques

All the preceding categories reduce the computational complexity of the BMME process at the expense of a suboptimal prediction quality. This category, however, reduces complexity without sacrificing prediction quality. It is interesting to note that algorithms in this category are usually based on ideas borrowed from the field of fast codebook search for vector quantization (VQ).

An example of the algorithms in this category is the *partial distortion elimination* (PDE) algorithm. Assume that during a full search, the minimum BDM calculated so far is $BDM(i_m, j_m)$ at search location (i_m, j_m). Then the BDM calculation of any subsequent search location (i, j) is stopped as soon as the accumulated distortion exceeds $BDM(i_m, j_m)$. This idea is very similar to the fast-search VQ method reported in Ref. 161. Clearly, initializing the search at a location with the lowest possible $BDM(i_m, j_m)$, achieves the highest possible reduction in computational complexity. As already shown in Section 4.6.7, the distribution of the best-match location is usually center-biased (i.e., the vector $(0, 0)$ has the highest probability, and longer vectors are less probable). Thus, the PDE algorithm is usually combined with a spiral-ordered search starting at the origin of the search space and going outward in a spiral fashion. This combination is employed, for example, in Telenor's H.263 codec [144].

Another algorithm in this category is the *successive elimination algorithm* (SEA) [162]. Again, this algorithm has similarities with the fast-search VQ algorithm reported in Ref. 163. The SEA algorithm is based on the triangular mathematical inequality given by

$$\left| \sum_{i=1}^{k} a_i \right| \leq \sum_{i=1}^{k} |a_i|, \tag{7.1}$$

where a_i are arbitrary real numbers. Extending this inequality to the SAD equation gives

$$\left| \sum_{(x,y)\in\mathscr{B}} |f_t(x, y)| - \sum_{(x,y)\in\mathscr{B}} |f_{t-\Delta t}(x - i, y - j)| \right|$$
$$\leq \sum_{(x,y)\in\mathscr{B}} |f_t(x, y) - f_{t-\Delta t}(x - i, y - j)|. \tag{7.2}$$

The first summation in this inequality is the sum norm of block $\mathcal{B}$ in the current frame, and this sum is denoted SN_t. The second summation, on the other hand, is the sum norm of a candidate block in the reference frame shifted by (i,j), and this sum is denoted $SN_{t-\Delta t}(i,j)$. The third summation is obviously the $SAD(i,j)$. Thus, for simplicity, Inequality (7.2) can be rewritten as

$$|SN_t - SN_{t-\Delta t}(i,j)| \leq SAD(i,j). \tag{7.3}$$

Now assume that during a full-search, the minimum SAD calculated so far is $SAD(i_m, j_m)$ at search location (i_m, j_m). A subsequent search location (i,j) is said to achieve better match only if $SAD(i,j) \leq SAD(i_m, j_m)$. Put in another way, and based on Inequality (7.3), a subsequent search location (i,j) is said to achieve better match only if $|SN_t - SN_{t-\Delta t}(i,j)| \leq SAD(i_m, j_m)$. In other words, a subsequent location (i,j) can be immediately skipped from the search if

$$|SN_t - SN_{t-\Delta t}(i,j)| \geq SAD(i_m, j_m). \tag{7.4}$$

Note that calculating the sum norms in this inequality has a reduced-complexity compared to calculating the $SAD(i,j)$ itself. For example, assume that $\mathcal{B}(x,y)$ is an $N \times N$ block with its top-left cornet at (x,y) and that the next block $\mathcal{B}(x+1,y)$ is the block obtained by moving one pel to the right. The two blocks are overlapping and they share $N-1$ columns. Once the sum norm, $SN(\mathcal{B}(x,y))$, of the first block is calculated, the sum norm, $SN(\mathcal{B}(x+1,y))$, of the next block in the line is obtained simply by substracting the sum norm of the first column of block $\mathcal{B}(x,y)$ and adding the sum norm of the last column in block $\mathcal{B}(x+1,y)$. A similar procedure can be used for calculating the sum norm of the next block in the column (i.e., when moving one pel down). Based on these ideas, a very fast method of calculating the sum norms is presented in Ref. 162.

A similar algorithm to the SEA has also been proposed in Ref. 164. Assume that $\mathcal{B}$ is partitioned into subsets $\mathcal{B}_n$ such that $\mathcal{B} = \bigcup_n \mathcal{B}_n$ and $\bigcap_n \mathcal{B}_n = \emptyset$. Then the triangular inequality becomes

$$\sum_n |SN_{t,n} - SN_{t-\Delta t,n}(i,j)| \leq SAD(i,j), \tag{7.5}$$

where $SN_{t-\Delta t,n}(i,j)$ is the sum norm over subset $\mathcal{B}_n$ of the candidate block in the reference frame shifted by (i,j). It can be shown that

$$\sum_n |SN_{t,n} - SN_{t-\Delta t,n}(i,j)| \geq |SN_t - SN_{t-\Delta t}(i,j)|. \tag{7.6}$$

Thus, a tighter bound is achieved when the partitioned case is used in Inequality (7.4) instead of the partitioned case. This tighter bound will result in faster rejection of more candidates and consequently will achieve higher speed-up ratios. However, the partitions must be chosen carefully to minimize the overhead calculations. In Ref. 164 two partitions are proposed. For an $N \times N$ block, the first partition produces N subsets, each being one of the N rows of the block, whereas the second partition produces N subsets, each being one of the N columns of the block. Again, this algorithm has similarities with the fast-search VQ algorithm presented in Ref. 47.

7.8   A Comparative Study

This section presents the results of a study comparing the categories of reduced-complexity motion estimation techniques discussed in Sections 7.3–7.7. The main aim of this study is to provide the reader with a feel of the relative performance of the discussed categories. Particular attention is given to the tradeoff between computational complexity and prediction quality.

In this study, one representative of each category was chosen. All simulated algorithms use 16×16 blocks, SAD as the distortion measure, ± 15 maximum displacement, full-pel accuracy, restricted motion vectors, and original reference frames. The simulated algorithms are:

FSA This is a full-search algorithm.

TDL This is the two-dimensional logarithmic search of Jain and Jain [54]. The algorithm is discussed in Section 7.3 and described in detail in the appendix, Section A.2.

SDM This algorithm uses a 4:1 subsampling of the matched blocks to reduce the complexity of calculating the distortion measure. The subsampling pattern used corresponds to pattern **A** described in Section 7.4. This pattern consists of all pels labeled **a** in Figure 7.2(a).

SMF This is the subsampled motion field algorithm of Liu and Zaccarin [157]. The algorithm is discussed in Section 7.5.

HME This is a three-level hierarchical motion estimation algorithm. The algorithm is described in Section 7.6 and illustrated using Figure 7.4.

PDE This is the partial distortion elimination algorithm described in Section 7.7. In order to reduce the overhead of logical operations, the condition to reject a given candidate is tested after accumulating the BDM of each row of the block (rather than after each pel of the block). The algorithm is supported with a spiral-ordered search starting at $(0,0)$ and going outward toward longer motion vectors.

Tables 7.2–7.4 present the results of testing these algorithms using the three test sequences AKIYO, FOREMAN, and TABLE TENNIS, with a frame skip of 1 (i.e., 30 frames/s for AKIYO and TABLE TENNIS and 25 frames/s for FOREMAN). All results are averages over sequences and refer to the luma components. Each table compares the algorithms in terms of prediction quality and computational complexity. The prediction quality is presented in terms of average luma PSNR in decibels. The difference in PSNR between each algorithm and the FSA is also shown.[4] The computational complexity is presented in terms of the average motion estimation time (in milliseconds) per frame.[5] Care

Table 7.2: Comparison between different fast block-matching algorithms when applied to QSIF AKIYO at 30 frames/s

	Prediction quality		Computational complexity	
	PSNR (dB)	ΔPSNR (dB)	ME Time (ms/frame)	Speed-up ratio
FSA	45.93	0.00	1013.87	1.00
PDE	45.93	0.00	48.49	20.91
SDM	45.93	0.00	278.25	3.64
SMF	45.93	0.00	511.51	1.98
TDL	45.93	0.00	26.82	37.80
HME	45.93	0.00	20.73	48.89

Table 7.3: Comparison between different fast block-matching algorithms when applied to QSIF FOREMAN at 25 frames/s

	Prediction quality		Computational complexity	
	PSNR (dB)	ΔPSNR (dB)	ME Time (ms/frame)	Speed-up ratio
FSA	32.20	0.00	1258.95	1.00
PDE	32.20	0.00	149.80	8.40
SDM	31.96	−0.24	346.72	3.63
SMF	31.91	−0.29	634.08	1.99
TDL	31.80	−0.40	34.76	36.22
HME	31.88	−0.32	25.73	48.92

---

[4] ΔPSNR = PSNR of fast algorithm − PSNR of FSA.

[5] Motion estimation times were obtained using the profiler of the Visual C++ 6.0 compiler run on a PC with a Pentium-III 700-MHz processor, 128 MB of RAM, and a Windows 98 operating system.

Table 7.4: Comparison between different fast block-matching algorithms when applied to QSIF TABLE TENNIS at 30 frames/s

	Prediction quality		Computational complexity	
	PSNR (dB)	ΔPSNR (dB)	ME time (ms/frame)	Speed-up ratio
FSA	32.17	0.00	1049.11	1.00
PDE	32.17	0.00	125.02	8.39
SDM	31.99	−0.18	287.73	3.65
SMF	31.44	−0.73	529.00	1.98
TDL	31.63	−0.54	28.66	36.61
HME	31.85	−0.32	21.62	48.54

should be taken when interpreting the results because the motion estimation time can vary with implementation and the underlying hardware platform. The speed-up ratio of each algorithm with reference to the FSA is also shown.[6]

As expected the FSA provides the best prediction quality, but at the expense of a high computational complexity.

The PDE algorithm provides an identical prediction quality to FSA, with a moderate speed-up ratio. Note that the computational complexity of PDE is highly dependent on the type of sequence and the motion content. For example, most blocks in the AKIYO sequence are stationary or quasi-stationary. Since PDE is initialized at $(0,0)$, this will lead to a very low starting minimum value $BDM(i_m, j_m)$. This will result in faster rejection of more candidates and, consequently, will lead to a relatively high speed-up ratio.

The SDM provides the next-best prediction quality. However, its 4:1 sub-sampling pattern limits its speed-up ratio to about 4. Similarly, the 2:1 field subsampling pattern of SMF limits its speed-up ratio to about 2. Note that the prediction quality of SMF is dependent on the amount of correlation between the motion vectors of neighboring blocks. This may explain the relatively high loss of prediction quality for the TABLE TENNIS sequence.

The TDL and HME algorithms provide the highest speed-up ratios, with moderate losses in prediction quality. In general, however, the HME algorithm outperforms the TDL algorithm in terms of both prediction quality and computational complexity.

---

[6]Speed-up $= \dfrac{\text{ME time for FSA}}{\text{ME time for fast algorithm}}$.

7.9 Discussion

Processing digital video requires a significant amount of computational power. This represents one of the main challenges for real-time mobile video communication, where processing power and battery life are scarce resources.

In this chapter, the computational complexity of a typical video codec was investigated. It was found that this complexity is due mainly to the motion estimation process. In fact, it was found that the computational complexity of this process is greater than that of all the remaining encoding steps combined. It was concluded, therefore, that reducing the complexity of this process is the best way to reduce the overall complexity of the codec. The chapter reviewed the main categories of reduced-complexity BMME techniques. The chapter then presented the results of a study comparing the different categories. It was found that hierarchical techniques and techniques based on a reduced set of motion vector candidates, in general, provide the highest reduction in computational complexity.

Chapter 8

The Simplex Minimization Search

8.1 Overview

As already discussed, one of the main requirements for mobile video communication is reduced complexity. In Chapter 7, it was shown that reducing the complexity of the motion estimation process is the best way to reduce the overall complexity of a video codec.

As detailed in Chapter 7 also, there are many techniques for reduced-complexity BMME. The most widely used approach is to use a reduced set of motion vector candidates. Algorithms in this category are usually based on a *unimodal error surface assumption*. In most cases, however, this assumption does not hold true, and such algorithms can easily get trapped in local minima, giving a suboptimal prediction quality. This chapter describes the design of a novel reduced-complexity BMME technique. Although this technique is based on using a reduced set of motion vector candidates, it is designed to be more robust against the local minimum problem.

BMME can be viewed as a *two-dimensional constrained minimization problem*. This problem can, therefore, be solved with reduced-complexity using a wealth of mature optimization techniques. This chapter solves the BMME optimization problem using the *simplex minimization* (SM) optimization method. The resulting solution is called the *simplex minimization search* (SMS). The initialization procedure, termination criterion, and constraints on the independent variables of the search are designed to take into account the basic properties of the BMME problem. This improves the prediction quality of the algorithm and, at the same time, increases its speed-up ratio.

In Chapter 6, it was concluded that one of the main drawbacks of multiple-reference motion-compensated prediction (MR-MCP) is the huge increase in computational complexity. To reduce complexity, this chapter extends the SMS algorithm to the multiple-reference case. Three different novel extensions (or

175

algorithms) are presented. They represent different degrees of compromise between prediction quality and computational complexity.

The rest of the chapter is organized as follows. Section 8.2 formulates BMME as a two-dimensional constrained optimization problem. The SM method and the reasons for choosing it to solve the BMME problem are described in Section 8.3. The design of the single-reference SMS algorithm is detailed in Section 8.4, and the results of testing it are presented in Section 8.5. Section 8.6 extends the SMS algorithm to the multiple-reference case. The chapter concludes with a discussion in Section 8.7.

Preliminary results of this chapter have appeared in Refs. 165, 166, 167, 168, and 169.

8.2 Block Matching: An Optimization Problem

8.2.1 Problem Formulation

As discussed in Chapter 4 (Section 4.6), in BMME the current frame, f_t, is usually partitioned into nonoverlapping blocks of $N \times N$ pels and the same motion vector is assigned to all pels within a block. The motion vector or displacement, $\mathbf{d} = [d_x, d_y]^T$, of a block is estimated by searching for the best-match block in a larger window of $(N + 2d_m) \times (N + 2d_m)$ pels centered at the same location in a reference frame, $f_{t-\Delta t}$, where d_m is the maximum allowed motion displacement. This process can be formulated as follows:

$$(d_x, d_y) = \arg \min_{i,j} \text{BDM}(i, j), \tag{8.1}$$

where $-d_m \le i, j \le +d_m$ and

$$\text{BDM}(i, j) = \sum_{y=1}^{N} \sum_{x=1}^{N} g[f_t(x, y) - f_{t-\Delta t}(x - i, y - j)]. \tag{8.2}$$

The BDM can be any positive function that measures the distortion between the block in the current frame and the candidate displaced block in the reference frame. Commonly used BDMs are the SSD, $g[\cdot] = (\cdot)^2$, and the SAD, $g[\cdot] = |\cdot|$.

Equations (8.1) and (8.2) clearly indicate that BMME is a *two-dimensional constrained optimization problem*. The two dimensions are the horizontal, i, and vertical, j, motion displacements, the function to be optimized (minimized in this case) is the BDM, and the independent variables, (i, j), are constrained within a limited range, $-d_m \le i, j \le +d_m$, and are usually evaluated to a certain accuracy, e.g., full- or half-pel accuracy.

An optimization problem can be thought of as a *search* process where the *function surface* is searched over a given *search space* to find its minimum (or maximum). This search is performed by examining the function value at a finite number of *search locations*. In BMME, the search space is the search window in the reference frame. Each candidate block within this window represents a search location, (i, j). This is the displacement between the block in the current frame and the candidate block in the reference frame. With full-pel accuracy, there are $(2d_m+1)^2$ possible search locations in the search space. The corresponding BDM values form the function surface. Since BDM is a distortion measure, this surface is also referred to as the *error surface*. The set of motion vectors assigned to the blocks of the frame form a *block-motion field*.

8.2.2   A Possible Solution

As shown in Section 8.2.1, BMME can be formulated as an optimization problem. This problem can, therefore, be solved, with reduced complexity, using a wealth of mature optimization methods.

There are few fast BMME algorithms that are based on optimization methods. For example, the TDL search of Jain and Jain [54] is an extension of the *1-D binary logarithmic search* [170], the OTS and CDS algorithms of Srinivasan and Rao [146] are based on the *conjugate directions* (CD) optimization method [171], and the GMS algorithm of Chow and Liu [148] is based on the *genetic algorithm* (GA) optimization method [172].

In a similar fashion, this chapter solves the BMME optimization problem using the *simplex minimization* (SM) optimization method [173]. The resulting solution is called the *simplex minimization search* (SMS).

Figure 8.1 shows the basic building blocks of any constrained optimization method. It can be seen that when trying to solve an optimization problem, there are *two* main design stages. The first, and probably the most important, stage is to choose a suitable optimization method. Section 8.3 describes the SM optimization method and outlines the reasons for choosing it to solve the BMME optimization problem. The second stage is to design a suitable initialization procedure, a termination criterion, and constraints on the independent variables of the search. For the SMS, this stage is detailed in Section 8.4.

8.3   The Simplex Minimization (SM) Optimization Method

8.3.1   Basic Algorithm

Simplex minimization (SM) is a *multidimensional unconstrained* optimization method that was introduced by Nelder and Mead in 1965 [173]. A *simplex* is a

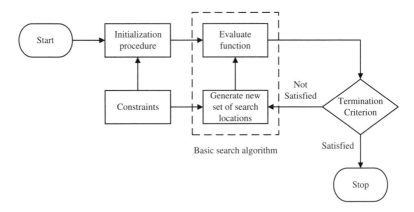

Figure 8.1: Basic building blocks of constrained optimization methods

geometrical figure that consists, in N dimensions, of $N+1$ vertices and all their interconnecting line segments, polygonal faces, etc. Thus, in two dimensions, a simplex is a triangle, whereas in three-dimensions it is a tetrahedron. A *non-degenerate simplex* is one that encloses a finite inner N-dimensional volume.

To minimize a function of N independent variables, the SM method must be initialized with $N+1$ points (or search locations) defining an *initial* nondegenerate simplex. The method then takes a series of steps: *reflecting*, *expanding*, or *contracting* the simplex from the point where the function value is largest, in an attempt to move it to a better point. Thus, the simplex is adapted to the local landscape of the function surface: expanded along inclined planes, reflected on encountering a valley at an angle, and contracted in the neighborhood of a minimum. This process continues until a termination criterion is satisfied. The SM method is described in more detail in Figure 8.2.

8.3.2 Simplex Minimization for BMME: Why?

The SM optimization method is an attractive choice for solving the BMME optimization problem for the following reasons:

1. Most fast BMME algorithms are based on a unimodal error surface assumption. As already shown (Property 4.6.7.3), this assumption does not hold true in most cases. For this reason, such algorithms are easily trapped in local minima, giving a suboptimal prediction quality. The SM method, however, is not based directly on this assumption.

2. Most fast BMME algorithms and optimization methods work by following the direction of the minimum distortion. The SM method,

1. The method is initialized with $N+1$ points, $\mathbf{p}_1,\ldots,\mathbf{p}_{N+1}$, defining an initial nondegenerate simplex where each point is in N dimensions, $\mathbf{p}_i = (p_{i,1},\ldots,p_{i,N})$. The function to be minimized, f, is then evaluated at those initial vertices to produce the function values $f_1 = f(\mathbf{p}_1),\ldots,f_{N+1} = f(\mathbf{p}_{N+1})$.

2. The highest, $f_h = \max_i f_i$, second highest, $f_s = \max_{i \neq h} f_i$, and lowest, $f_l = \min_i f_i$, function values are determined and the corresponding vertices are marked as $\mathbf{p}_h$, $\mathbf{p}_s$, and $\mathbf{p}_l$, respectively. The centroid, $\mathbf{p}_m$, of the simplex with the highest point removed is then evaluated using

$$\mathbf{p}_m = \frac{1}{N} \sum_{i \neq h} \mathbf{p}_i. \qquad (8.3)$$

3. It would seem reasonable to move away from $\mathbf{p}_h$. Thus, the simplex is *reflected* from its highest point about its centroid using

$$\mathbf{p}_r = -\alpha \mathbf{p}_h + (1 + \alpha)\mathbf{p}_m, \qquad (8.4)$$

where $\mathbf{p}_r$ is the reflected point and $\alpha \geq 0$ is the *reflection coefficient*. The function is then evaluated at this new reflected point, giving $f_r = f(\mathbf{p}_r)$.

4. IF ($f_r < f_l$), then reflection has produced the lowest function value. Therefore, the direction from $\mathbf{p}_m$ to $\mathbf{p}_r$ seems to be a good one to move along. Thus, the simplex is *expanded* in this direction using

$$\mathbf{p}_e = \gamma \mathbf{p}_r + (1 - \gamma)\mathbf{p}_m, \qquad (8.5)$$

where $\mathbf{p}_e$ is the expanded point and $\gamma \geq 1$ is the *expansion coefficient*. The function is then evaluated at this new expanded point, giving $f_e = f(\mathbf{p}_e)$. There are now two possible cases:

   (a) IF ($f_e < f_l$), then the expansion step was in the right direction. Thus, $\mathbf{p}_h$ is removed from the simplex and replaced by $\mathbf{p}_e$. The search then proceeds to step 8 to test for convergence.

   (b) ELSE it seems that the expansion step moved too far in the direction from $\mathbf{p}_m$ to $\mathbf{p}_r$. Thus, $\mathbf{p}_e$ is abandoned. Since $\mathbf{p}_r$ is already known to produce an improvement, $\mathbf{p}_h$ is removed from the simplex and replaced by $\mathbf{p}_r$. The search then proceeds to step 8 to test for convergence.

5. ELSE IF ($f_r > f_l$ AND $f_r < f_s$), then the reflected point is an improvement over the worst two points of the simplex. Thus, $\mathbf{p}_h$ is removed from the simplex and replaced by $\mathbf{p}_r$. The search then proceeds to step 8 to test for convergence.

6. ELSE IF ($f_r > f_i$, for all $i \neq h$), then there are two possible cases:

   (a) IF ($f_r > f_h$), then the search proceeds directly to the contraction step (step 7).

   (b) ELSE $\mathbf{p}_h$ is first removed from the simplex and replaced by $\mathbf{p}_r$ and then the search proceeds to the contraction step (step 7).

Figure 8.2: Simplex method for function minimization

7. It seems that the reflection step moved too far in the direction from $\mathbf{p}_h$ to $\mathbf{p}_m$. This is rectified by *contracting* the simplex from its highest point toward its centroid using

$$\mathbf{p}_c = \beta\mathbf{p}_h + (1 - \beta)\mathbf{p}_m, \tag{8.6}$$

where $\mathbf{p}_c$ is the contracted point and $0 \leq \beta \leq 1$ is the *contraction coefficient*. The function is then evaluated at the new contracted point, giving $f_c = f(\mathbf{p}_c)$. There are now two possible cases:

(a) IF ($f_c < f_h$), then contraction has produced a better point. Thus, $\mathbf{p}_h$ is removed from the simplex and replaced by $\mathbf{p}_c$. The search then proceeds to step 8 to test for convergence.

(b) ELSE it would appear that all the efforts to move the highest point to a better location has failed. All the vertices are, therefore, pulled toward the lowest point using

$$\mathbf{p}_i = \frac{\mathbf{p}_i + \mathbf{p}_l}{2}, \quad \text{for all } i \tag{8.7}$$

8. Convergence is tested. IF the convergence criterion is satisfied, then the search is stopped. ELSE the search goes back to step 2.

Figure 8.2: *Continued.*

however, works by moving the point where the function value is largest in different directions using reflection, expansion, and contraction. Thus, it explores directions other than that of the minimum distortion. This makes the method more resilient to the local minimum problem.

3. As shown in Figure 8.1, a very important process in any optimization method is the generation of a new set of search locations for the next iteration. The performance and complexity of any method is highly dependent on this process. The simplest approach is to use a predetermined uniform distribution of search locations. This approach is adopted by most fast BMME algorithms (see the Appendix). There are, however, more complex approaches, like the use of crossover and mutation operators in genetic algorithms or the use of gradients in gradient-descent algorithms. The SM method is a compromise between the two extremes. It uses very simple equations for reflection (8.4), expansion (8.5) and contraction (8.6), as shown in Figure 8.2. As will be shown later, a suitable choice of the coefficients, (α, β, γ), can further reduce the complexity of such equations.

8.4 The Simplex Minimization Search (SMS)

Having decided on the optimization method to be used (the SM optimization method in this case), the second stage is to design a suitable initialization procedure, a termination criterion, and constraints on the independent variables of the search. The performance of an optimization method can be greatly improved if this design stage exploits *a priori* knowledge of the problem at hand. For example, the basic properties of the BMME problem can be exploited to avoid local minima and initialize the search at a location close to the global minimum. This improves the prediction quality and at the same time increases the speed-up ratio.

Although the TDL, OTS, CDS, and GMS algorithms are all based on good optimization methods, they do not take into account the basic properties of the BMME problem. As a result, such algorithms can either get trapped in local minima, resulting in suboptimal prediction quality, or lead to a relatively small speed-up ratio. In the *simplex minimization search* (SMS) algorithm, however, the initialization procedure, termination criterion, and constraints on the independent variables of the search are designed to exploit the basic properties of the BMME problem. This is described in more detail in the following subsections.

8.4.1 Initialization Procedure

Block-matching motion estimation is a two-dimensional problem. As already mentioned, a simplex, in two-dimensions, is a *triangle*. Thus, *three* points need to be chosen to define the initial nondegenerate simplex. As is shown later, the performance of the SM method is highly dependent on the choice of these points. The following initialization procedure is used.

According to Property 4.6.7.1, the vector $(0,0)$ has the highest probability of occurrence within the block-motion field. One of the initial three points is therefore set to $(0,0)$. In addition, Property 4.6.7.2 states that there is a high correlation between the motion vectors of adjacent blocks. In fact, most video coding standards take advantage of this property by predictively coding the motion vectors. To exploit this property, and to match the motion estimation process to the motion coding process, the other two points of the initial simplex are set to the motion vectors of the blocks *above* and to the *left* of the current block. If such neighboring vectors are not available, as in border blocks, they are set to $(0,0)$.

Note that this procedure does not guarantee to produce a nondegenerate initial simplex. For example, if two points are identical, then the simplex is degenerate. In this case, a *local* search is applied to find other candidates. The BDM is first evaluated at the points chosen by the foregoing procedure.

Let $\mathbf{p}_m = (m_x, m_y)$ be the point that yields the smallest BDM, then the BDM is also evaluated at its eight nearest neighbors, $(m_x, m_y \pm a)$, $(m_x \pm a, m_y)$ and $(m_x \pm a, m_y \pm a)$, where a is the accuracy of the search, e.g., $a = 1$ for full-pel accuracy. At this stage, all points (including those from the initial procedure) are arranged in ascending order according to their BDMs and the first three are chosen to form the initial simplex. If this is still a degenerate one, then the appropriate point is dropped and replaced by the next one in the list. This is repeated until a nondegenerate simplex is formed.

Once a nondegenerate initial simplex is formed, the search proceeds as shown in Figure 8.2, subject to the constraints outlined in Section 8.4.2, and is terminated when the criterion described in Section 8.4.3 is satisfied.

8.4.2   Constraints on the Independent Variables

The SM method assumes continuous unconstrained independent variables. However, when applied to the constrained minimization problem of BMME, two constraints have to be imposed. Firstly, the vertices of the simplex must always be set to the required accuracy before any BDM evaluation can take place. For example, if full-pel accuracy is assumed, then any point produced by reflection, expansion, or contraction must be rounded to the nearest integer value. Secondly, the vertices of the simplex must always be kept within the search window. Any point produced by reflection, expansion, or contraction must be set to the closest point within the range $-d_m \leq i, j \leq +d_m$ before any BDM evaluation can take place. This constraint is more efficient than other possible constraints, like, for example, assigning a large function value to the vertex outside the search window.

8.4.3   Termination Criterion

There are many possible ways to terminate optimization methods. One of the most widely used approaches is to terminate the search if the current minimum function value is below some threshold. In the SM case, another approach is to terminate the search if the fractional range from the highest, in terms of function value, to the lowest vertices of the simplex is below some threshold [174].

According to Property 4.6.7.4, the function value at the global minimum is unpredictable. Thus, if the preceding termination criteria are used, then the threshold needs to be adjusted from one sequence to another, from one frame to another, and even from one block to another. This makes such criteria unsuitable for BMME. A more suitable criterion is as follows. Let $\mathbf{p}_h = (h_x, h_y)$, $\mathbf{p}_s = (s_x, s_y)$, and $\mathbf{p}_l = (l_x, l_y)$ be the vertices of the simplex where the BDM is highest, second highest, and lowest, respectively. The search is terminated if the following condition is satisfied:

$$(|h_x - l_x| \leq a) \wedge (|h_y - l_y| \leq a) \wedge (|s_x - l_x| \leq a) \wedge (|s_y - l_y| \leq a), \qquad (8.8)$$

where a is the search accuracy and $\wedge$ is the logical AND operator. In other words, the search is terminated *if the two highest (in terms of BDM value) vertices of the simplex become neighbors to the lowest vertex.* This criterion was derived from the way the SM method works. As shown in Figure 8.2, when the method converges to a minimum, the contraction operation starts pulling all the vertices toward the minimum vertex. The main advantage of this criterion is that it does not depend on a threshold.

8.4.4   Motion Vector Refinement

The main disadvantage of the preceding termination criterion is that it is not based directly on the function to be minimized, i.e., the BDM. As a result, the search may sometimes converge to a suboptimal point. Experimental results show that in most cases this suboptimal point is in the neighborhood of the global minimum. An extra step is therefore added to the search in which the motion vector produced by SM is *refined* by searching its eight nearest neighbors. Note that this does not significantly increase the complexity of the search, because most of those neighbors have already been searched.

8.5   Simulation Results

8.5.1   Results Within an Isolated Test Environment

In this set of simulations, motion is estimated and compensated using original reference frames. In effect, this is equivalent to *lossless* DFD coding. This is particularly important for a fair comparison between different algorithms on a frame-by-frame basis, since poor prediction of one frame does not propagate to, and affect the prediction of, the next frame. Hereafter, the term *isolated test environment* will be used to refer to this test condition.

All results in this subsection were generated using blocks of 16×16 pels, a maximum allowed displacement of ± 15 pels, SAD as the distortion measure, restricted motion vectors, and full-pel accuracy. Motion vectors were coded predictively using the prediction method and the VLC table of the H.263 standard. All quoted results refer to the luma components of sequences.

8.5.1.1 Choice of Coefficients

Before evaluating the performance of the SMS algorithm, suitable values for the reflection, α, contraction, β, and expansion, γ, coefficients need to be chosen. Figures 8.3, 8.4, and 8.5 show the performance of the SMS algorithm with different values of α, β, and γ, respectively. The figures indicate that the performance of the SMS algorithm is not very sensitive to the choice of these coefficients. This may be due to the good performance of the initialization procedure and termination criterion. In general, however, the values $\alpha = 1$, $\beta = \frac{1}{2}$, and $\gamma = 2$ provide the best compromise between computational complexity and prediction quality. In addition, this particular set of coefficients reduces the complexity of the SM transformation equations, Equations (8.4),

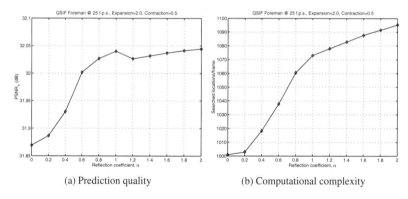

(a) Prediction quality   (b) Computational complexity

Figure 8.3: Performance of SMS with different values of the reflection coefficient α

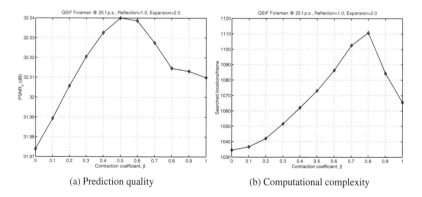

(a) Prediction quality   (b) Computational complexity

Figure 8.4: Performance of SMS with different values of the contraction coefficient β

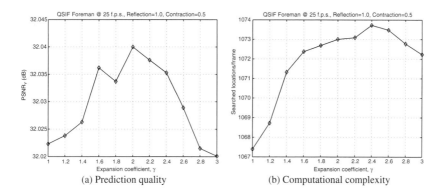

(a) Prediction quality   (b) Computational complexity

Figure 8.5: Performance of SMS with different values of the expansion coefficient γ

(8.5), and (8.6) in Figure 8.2, because multiplications and divisions in this case can be performed using shift operations.

8.5.1.2 Initialization, Termination, and Refinement Tests

In order to justify different parts of the SMS algorithm, the following tests were performed.

1. **Initialization Test**: Two initialization procedures were tested:

   (a) *Random Initialization*: Two of the vertices of the initial simplex are generated randomly within the search window, whereas the third vertex is always set to $(0,0)$.

   (b) *Proposed Initialization*: This is the initialization procedure described in Section 8.4.1.

2. **Termination Test**: Two termination criteria were tested.

   (a) *Threshold Termination*: The search is terminated when the current minimum BDM value is below a threshold. The threshold was set to 768, which corresponds to an average SAD/pel of 3 ($16 \times 16 \times 3$). As already discussed, a fixed threshold is not suitable for the BMME problem. Such a threshold does not guarantee convergence because the global minimum BDM value may in some cases be above the threshold. The threshold condition must therefore be supported by another condition to guarantee termination. In this test, the search is also terminated if the number of iterations exceeds 10.

   (b) *Proposed Termination*: This is the termination criterion described in Section 8.4.3.

Table 8.1: Initialization, termination, and refinement tests

	AKIYO		FOREMAN		TABLE TENNIS	
	PSNR (dB)	Locations	PSNR (dB)	Locations	PSNR (dB)	Locations
Initialization test:						
Random	45.93	1,634	31.32	1,890	31.28	1,647
Proposed	45.93	684	32.04	1,073	31.71	831
Termination test:						
Threshold	45.93	904	32.06	1,396	31.73	1,082
Proposed	45.93	684	32.04	1,073	31.71	831
Refinement test:						
No refinement	45.93	683	31.97	999	31.53	794
Proposed	45.93	684	32.04	1,073	31.71	831

3. **Refinement Test**: Two cases were tested.

(a) *Proposed Refinement*: The motion vector produced by SM is refined by searching its eight nearest neighbors, as described in Section 8.4.4.

(b) *No Refinement*: No refinement is performed.

Table 8.1 summarizes the results of the preceding tests. The results are averaged over each sequence with a frame skip of 1. Prediction quality is given in terms of average luma PSNR (dB), and computational complexity is given in terms of average searched locations per frame. The results clearly justify the use of the proposed initialization procedure, termination criterion, and refinement step.

8.5.1.3   Performance Evaluation

In addition to the SMS algorithm, five BMME algorithms were simulated: the full-search (FS) algorithm, the two-dimensional logarithmic search (TDL) [54], the cross-search algorithm (CSA) [147], the one-at-a-time search (OTS) [146], and the N-steps search (NSS), which is the general form[1] of the three-steps search (TSS) [145]. In this case the number of steps in the NSS search is set to $N = 4$ to give a maximum displacement of ± 15 pels. A detailed description of these fast BMME algorithms is given in the Appendix.

---

[1] The three-steps search starts with ± 4 pels in the first step, then ± 2 pels in the second step, and ± 1 pel in the third step. This gives a maximum allowed displacement of $\pm 4 \pm 2 \pm 1 = \pm 7$ pels. For larger search windows the number of steps must be increased. This is called the N-steps search. For example, when $N = 4$, the search has 4 steps and the first step starts with ± 8 pels, giving a maximum allowed displacement of ± 15 pels.

Table 8.2: Comparison between different block-matching algorithms in terms of prediction quality

	AKIYO			FOREMAN			TABLE TENNIS		
	PSNR	ΔPSNR	% Global	PSNR	ΔPSNR	% Global	PSNR	ΔPSNR	% Global
FS	45.93	0.00	100.00	32.20	0.00	100.00	32.17	0.00	100.00
SMS	**45.93**	**0.00**	**100.00**	**32.04**	**−0.16**	**94.31**	**31.71**	**−0.46**	**95.80**
NSS	45.93	0.00	100.00	31.74	−0.46	87.25	31.50	−0.67	92.74
TDL	45.93	0.00	100.00	31.81	−0.39	88.92	31.63	−0.54	93.39
CSA	45.91	−0.02	99.86	30.95	−1.25	60.11	30.93	−1.24	81.23
OTS	45.93	0.00	100.00	31.23	−0.97	76.35	31.23	−0.94	91.60

Table 8.3: Comparison between different blockmatching algorithms in terms of computational complexity

	AKIYO		FOREMAN		TABLE TENNIS	
	Locations	Speed-up	Locations	Speed-up	Locations	Speed-up
FS	65,621	–	77,439	–	65,621	–
SMS	**684**	**96**	**1,073**	**72**	**831**	**79**
NSS	2,464	27	2,823	27	2,473	27
TDL	1,310	50	1,638	47	1,362	48
CSA	115	571	920	84	461	142
OTS	402	163	604	128	448	146

Tables 8.2, 8.3, and 8.4 compare the performance of the simulated BMME algorithms. All results are averages over sequences with a frame skip of 1. Table 8.2 compares the prediction quality in terms of average luma PSNR in decibels. The difference in PSNR between each algorithm and the FS algorithm is also shown.[2] The table also shows the average percentage of finding the global minimum. Table 8.3, on the other hand, compares the computational complexity in terms of average searched locations per frame. It also shows the speed-up ratio[3] of each algorithm with reference to the FS algorithm. Table 8.4 shows the motion overhead generated by each algorithm and the difference between this overhead and that produced by the FS algorithm.[4]

As expected, the FS algorithm provides the best prediction quality, but at the expense of a very high computational complexity. The fast BMME algorithms in this simulation can be split into three difference performance classes. In the first class, the CSA and the OTS algorithms provide the highest

---

[2] ΔPSNR = PSNR of fast algorithm − PSNR of FS algorithm.

[3] Speed-up = $\dfrac{\text{Searched locations for FS algorithm}}{\text{Searched locations for fast algorithm}}$.

[4] ΔBits = Motion bits of fast algorithm − Motion bits of FS algorithm.

Table 8.4: Comparison between different block-matching algorithms in terms of motion overhead

	Akiyo		Foreman		Table Tennis	
	Motion bits	ΔBits	Motion bits	ΔBits	Motion bits	ΔBits
FS	177	0	388	0	279	0
SMS	**177**	**0**	**358**	**−30**	**247**	**−32**
NSS	177	0	457	+69	290	+11
TDL	177	0	394	+6	269	−10
CSA	177	0	461	+73	281	+2
OTS	177	0	388	0	246	−33

speed-up ratios, but their prediction quality deteriorates for sequences with medium to high movement content. In the second class, the NSS and the TDL algorithms provide better prediction quality than CSA and OTS, but at the expense of a higher computational complexity. In the third class, the SMS algorithm provides the best compromise between prediction quality and computational complexity. Its prediction quality is the closest to that of FS and yet its computational complexity is between those of the other two classes. Note that the SMS algorithm achieves the highest percentage of finding the global minimum. This clearly indicates that the SMS algorithm is the most resilient to the local minimum problem. Note also that the SMS algorithm adapts better to the movement content of sequences. Thus, for low-movement sequences it uses fewer locations and for high-movement sequences it uses more locations. In addition, because the motion estimation process is matched to the motion coding process (through the initialization procedure), the SMS algorithm has the lowest motion overhead.

One of the disadvantages of fast BMME algorithms is that their prediction quality deteriorates for higher amounts of motion and larger search windows (as, for example, in HDTV applications). This is clear from Table 8.2 when moving from Akiyo to Foreman and Table Tennis. To investigate this effect further, the Foreman sequence was temporally subsampled to $25, 12.5, 8.33$, and 6.25 frames/s (this corresponds to frame skips of 1, 2, 3, and 4, respectively). The corresponding maximum allowed displacements, d_m, were set to ± 7, ± 15, ± 31, and ± 63 pels, respectively. Figure 8.6 shows the results of this simulation. It is immediately evident that the SMS algorithm is the most robust fast algorithm to the above effect, and yet it has the second-lowest computational complexity.

8.5.2   Results Within an H.263-like Codec

The SMS algorithm along with the other five BMME algorithms have also been tested within a hybrid H.263-like codec. As in previous simulations,

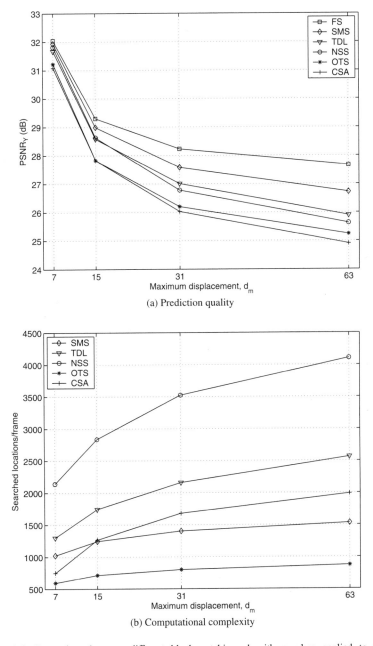

(a) Prediction quality

(b) Computational complexity

Figure 8.6: Comparison between different block-matching algorithms when applied to QSIF FOREMAN with maximum displacements of 7, 15, 31, and 63 and corresponding frame rates of 25, 12.5, 8.33, and 6.25 frames/s, respectively

motion was estimated using macroblocks of 16×16 pels, a maximum allowed displacement of ± 15 pels, SAD as the distortion measure, restricted motion vectors, and full-pel accuracy. In this case, however, motion vectors were predictively encoded using the median prediction method and the VLC table of the H.263 standard. In addition, motion was estimated and compensated using reconstructed reference frames rather than original frames. Both, the frame signal (in case of INTRA) and the DFD signal (in case of INTER) were transform encoded according to the H.263 standard. To generate a range of bit rates, the quantization parameter QP was varied over the range 5–30 in steps of 5. This means that each algorithm was used to encode a given sequence six times. Each time, QP was held constant over the whole sequence (i.e., no rate control was used). The first frame was always INTRA encoded, and all other frames were INTER encoded. No INTRA/INTER switching was allowed at the macroblock level. The INTRA bits were included in the bit-rate calculations, and no header bits were generated. All quoted results refer to the luma components of sequences.

Figures 8.7 and 8.8 show examples of the rate-distortion (R-D) performance of the SMS algorithm and compare it to that of the other five BMME algorithms. Figure 8.7 shows the results for the FOREMAN sequence with frame rates of 25 frames/s and 8.33 frames/s, whereas Figure 8.8 shows the results for the AKIYO and TABLE TENNIS sequences with frame rates of 10 frames/s and 15 frames/s, respectively. Both figures confirm the superior R-D performance of the SMS algorithm compared to other fast BMME algorithms.

The superior performance of the SMS algorithm is also shown on a frame-by-frame basis in Figure 8.9. This figure shows the performance for the FOREMAN sequence at 8.33 frames/s with a quantization parameter of QP = 10. For clarity, the figure shows only the performance of the FS, SMS, NSS, and OTS algorithms. As can be seen, the SMS algorithm provides the closest prediction quality (Figure 8.9(a)) to the FS algorithm. This results in the use of fewer bits for the DFD signal (Figure 8.9(c)). In addition, the initialization procedure results in less motion overhead (Figure 8.9(d)). The reduced number of DFD bits and motion bits results in a reduced overall bit rate (Figure 8.9(e)). This is all achieved at a reduced computational complexity (Figure 8.9(b)).

8.5.3   Results Within an MPEG-4 Codec

In a collaborative work, the SMS algorithm has also been tested within an MPEG-4 codec. The results in this subsection are reproduced, as is, from Ref. 175.[5]

---

[5]The authors would like to thank Mr. Oliver Sohm for incorporating SMS within MPEG-4 and providing the results.

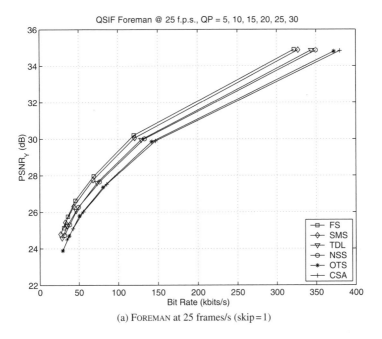

(a) FOREMAN at 25 frames/s (skip = 1)

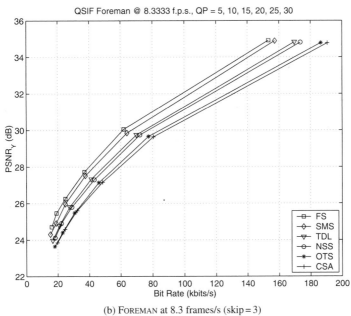

(b) FOREMAN at 8.3 frames/s (skip = 3)

Figure 8.7: R-D performance of different block-matching algorithms when applied to QSIF FOREMAN

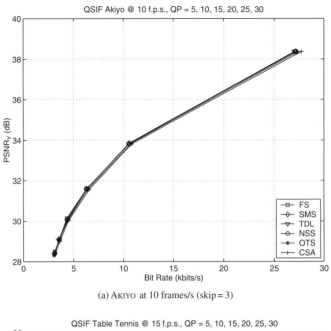

(a) AKIYO at 10 frames/s (skip = 3)

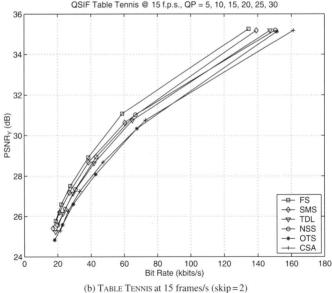

(b) TABLE TENNIS at 15 frames/s (skip = 2)

Figure 8.8: R-D performance of different block-matching algorithms when applied to QSIF AKIYO and QSIF TABLE TENNIS

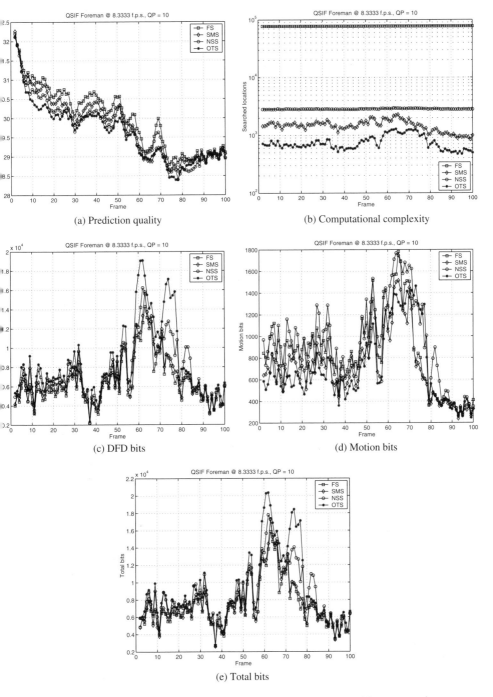

Figure 8.9: Comparison between different block-matching algorithms.

They are provided here to show the performance of the SMS algorithm within an object-based video codec.

Before proceeding to present the results, a description of object-based motion estimation in the MPEG-4 verification model [176] is in order. To account for arbitrarily shaped objects, the standard block-matching algorithm is extended to *polygon matching*. Macroblock-based repetitive padding is used for the reference visual object plane (VOP). In other words, macroblocks that lie on the VOP boundary are padded so that pels from inside the VOP are extrapolated to the outside. For each 16×16 macroblock in the current VOP, full-pel full search is used to find the motion vector that minimizes the SAD. The SAD of the motion vector $(0,0)$ is reduced by a preset threshold to favor this vector. A reduced search of ± 2 pels centered around the 16×16 motion vector is used to find one motion vector for each of the four 8×8 blocks within the MB. A decision is then made whether to use one motion vector or four motion vectors per MB. A decision is also made whether to encode the MB in INTRA or INTER mode. If INTER mode is chosen, the 16×16 (or the four 8×8) vector(s) is/are refined to half-pel accuracy using a reduced $\pm 1/2$-pel search centered around the full-pel vector. Motion vectors are restricted within the bounding box of the VOP unless the unrestricted mode is chosen. In this mode, the reference VOP is extended by repetitive padding in all directions by the number of pels which equals the search range. Overlapped motion compensation is similar to that of H.263.

In this set of simulations, four algorithms were tested: FS, SMS, NSS, and diamond search (DS) [149, 150, 151] (which is adopted in the MPEG-4 verification model [176]). To ensure that the global minimum is found, the threshold that favors the $(0,0)$ vector in the FS algorithm was set to zero. The four algorithms were used only for the full-pel search. All other operations (e.g., 8×8 ME, half-pel refinement) remained the same. Original reference VOPs were used instead of reconstructed VOPs. The unrestricted motion vector mode was switched on. Table 8.5 gives more details about the test conditions and the test sequences.

Table 8.6 shows the prediction quality in terms of mean absolute error per pel (MAE/pel),[6] whereas Table 8.7 shows the computational complexity in terms of average searched locations per macroblock (locations/MB). Again, the superior performance of the SMS algorithm is evident. Compared to NSS and DS, the SMS algorithm provides the closest MAE/pel to that of FS, and yet it has the least number of searched locations/MB.

---

[6]The MAE/pel measure was calculated as follows. The minimum SADs over the whole VOP were summed and then divided by the number of opaque pels in the VOP. The minimum SADs in this case are those produced by the full-pel search.

Table 8.5: Test sequences and conditions for the MPEG-4 results. Reproduced from Ref. 175

Sequence	Format	Class	Objects	Distance	Displacement
Bream	CIF, 352 × 288, 30 Hz, 300 frames	E	VO0: Background VO1: Fish	2 (15 f.p.s.)	−16...15
Coast Guard	QCIF, 176 × 144, 30 Hz, 270 frames	B	VO0: Water VO1: Small boat VO2: Big boat VO3: River bank	3 (10 f.p.s.)	−16...15
Container ship	SIF, 352 × 240, 30 Hz, 161 frames	A	VO0: Water VO1: Ship VO2: Small boat VO3: Land (fg) VO4: Sky+Land (bg) VO5: Flag	4 (7.5 f.p.s.)	−16...15
News	QCIF, 176 × 144, 30 Hz, 300 frames	B	VO0: Background VO1: Dancers VO2: News readers VO3: Text	3 (10 f.p.s.)	−16...15
Stefan	CIF, 352 × 288, 30 Hz, 300 frames	C	VO0: Stefan	1 (30 f.p.s.)	−16...15

Table 8.6: Prediction quality within MPEG-4 in terms of MAE/pel. Reproduced from Ref. 175

Sequence	Object	FS	SMS	DS	NSS
Bream	VO1: Fish	6.277	6.533	8.416	9.709
Coast Guard	VO0: Water	3.797	3.847	3.993	4.014
	VO1: Small boat	5.197	5.374	5.692	5.543
	VO2: Big boat	4.696	4.898	5.096	5.071
	VO3: River bank	4.591	4.636	4.885	6.523
Container ship	VO0: Water	2.287	2.288	2.357	2.358
	VO1: Ship	2.069	2.069	2.082	2.113
	VO2: Small boat	2.154	2.159	2.166	2.181
	VO3: Land (fg)	0.831	0.831	0.831	0.831
	VO4: Sky+Land (bg)	0.792	0.801	0.839	0.843
	VO5: Flag	15.828	16.066	16.105	16.131
News	VO0: Background	0.060	0.061	0.061	0.061
	VO1: Dancers	5.568	5.773	5.860	5.852
	VO2: News readers	1.153	1.154	1.159	1.158
	VO3: Text	0.092	0.092	0.092	0.092
Stefan	VO0: Stefan	8.200	8.662	9.346	9.430
Average		3.975	4.078	4.311	4.494
Relative to FS		100.0%	102.6%	108.5%	113.1%

Table 8.7: Computational complexity within MPEG-4 in terms of searched locations/MB. Reproduced from Ref. 175

Sequence	Object	FS	SMS	DS	NSS
Bream	VO1: Fish	1,017.2	16.7	21.4	33.0
Coast Guard	VO0: Water	1,021.8	16.4	18.0	33.0
	VO1: Small boat	1,004.7	16.8	17.5	32.9
	VO2: Big boat	1,010.0	17.7	19.1	33.0
	VO3: River bank	1,020.5	13.1	19.2	32.9
Container ship	VO0: Water	1,023.4	12.7	13.1	33.0
	VO1: Ship	1,023.7	11.4	13.5	33.0
	VO2: Small boat	1,014.2	14.7	15.9	32.9
	VO3: Land (fg)	1,024.0	9.4	13.0	33.0
	VO4: Sky+Land (bg)	1,024.0	13.9	13.1	33.0
	VO5: Flag	1,012.3	19.1	15.6	33.0
News	VO0: Background	1,024.0	9.3	13.0	33.0
	VO1: Dancers	1,024.0	15.4	16.3	33.0
	VO2: News readers	1,022.9	9.8	13.1	33.0
	VO3: Text	1,024.0	9.0	13.0	33.1
Stefan	VO0: Stefan	1,002.4	21.8	22.2	32.9
Minimum		1,002.4	9.0	13.0	32.9
Maximum		1,024.0	21.8	22.2	33.1
Average		1,018.3	14.2	16.1	33.0

8.6 Simplex Minimization for Multiple-Reference Motion Estimation

As already discussed, MR-MCP achieves significant prediction gains, but at the expense of a significant increase in computational complexity. This is illustrated in Figure 8.10 for the FOREMAN sequence at 8.33 frames/s. This figure was generated using the same simulation conditions described in Section 6.3.2. Figure 8.10(a) shows the prediction quality (in terms of $PSNR_Y$ in decibels) as a function of multiframe memory size (in frames), whereas Figure 8.10(b) shows the computational complexity (in terms of searched locations/frame). It is clear that increasing the memory size M increases the prediction quality. This is, however, at the expense of a linear increase in computational complexity. The aim of this section is to design fast long-term memory block-matching algorithms that can reduce computational complexity but at the same time maintain the prediction gain of multiple-reference motion estimation.

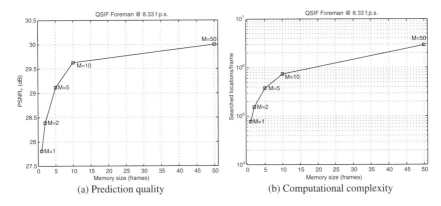

Figure 8.10: Performance of LTM-MCP as a function of memory size for QSIF FOREMAN at 8.33 frames/s

8.6.1 Multiple-Reference SMS Algorithms

This section extends the SMS algorithm to the multiple-reference case. As detailed in Section 8.4, the design of the SMS algorithm was based on some important properties of the block-motion fields of typical video sequences. In particular, the design was based on Properties 4.6.7.1 and 4.6.7.2 of the single-reference block-motion field. The two properties are the center-biased distribution of the field and the high correlation between adjacent motion vectors, respectively. The results of the investigation in Section 6.3.1 indicate that the two properties still hold true in the multiple-reference case (Properties 6.3.1.1 and 6.3.1.3). Thus, the efficient performance of the SMS algorithm can be extended to the multiple-reference case without the need for a major redesign. Three different extensions (or algorithms) are described in what follows.

MR-SMS This is a direct extension of SMS. For each block in the current frame, the single-reference SMS algorithm is used to individually search each frame in the multiframe memory and produce a best-match block from that frame. The overall best-match is then chosen from this set of M blocks.

MR-FS/SMS This is the same as MR-SMS, but the most recent reference frame in memory (i.e., the frame for which $d_t = 0$) is searched using full search instead of SMS. Giving more importance to searching this frame is motivated by Property 6.3.1.2, which states that the most recent reference frame has the highest probability of selection.

MR-3DSM The single-reference SMS algorithm is based on a two-dimensional version of the simplex minimization (SM) optimization method

(Section 8.3). Algorithm MR-3DSM, however, is based on a *three-dimensional* version ($N = 3$ in Figure 8.2). A 3-D version of SM must be initialized with *four* locations defining an initial simplex in the search space. For the MR-3DSM algorithm, this is achieved as follows. For each block in the current frame, the initialization procedure described in Section 8.4.1 is applied individually to each frame in the multiframe memory. This will generate three initial vertices from each frame. The best *four* vertices, in terms of BDM value, are selected from this set of $3M$ vertices. A procedure similar to that described in Section 8.4.1 is used to ensure that the four vertices form a nondegenerate simplex. This simplex is used to initialize the 3-D version of SM, where the third dimension here is the temporal displacement. The same criterion described in Section 8.4.3 is used to terminate the algorithm, with the added condition that *the four vertices of the final simplex must have the same temporal displacement.*

8.6.2 Simulation Results

The multiple-reference SMS algorithms were tested using the luma components of the three QSIF sequences AKIYO, FOREMAN, and TABLE TENNIS with full-pel accuracy, blocks of 16×16 pels, a maximum allowed displacement of ± 15 pels, SAD as the distortion measure, restricted motion vectors, and original reference frames. In addition to the multiple-reference SMS algorithms, the single-reference full-search (SR-FS) and the multiple-reference full-search (MR-FS) algorithms were also simulated. For the multiple-reference algorithms, sliding-window control was used to maintain a long-term memory of size $M = 50$ frames.

Tables 8.8 and 8.9 compare the performance of the simulated algorithms. All results are averages over sequences with a frame skip of 1. Table 8.8 compares the prediction quality in terms of average luma PSNR in decibels. The difference[7] in PSNR between each algorithm and the MR-FS algorithm is also shown. Table 8.9, on the other hand, compares the computational complexity in terms of average searched locations per frame. It also shows the speed-up ratio[8] of each algorithm with reference to the MR-FS algorithm.

It is immediately evident that the multiple-reference SMS algorithms provide significant reductions in computational complexity compared to the MR-FS algorithm. The SMS algorithms represent different degrees of compromise between prediction quality and computational complexity. At one extreme is

---

[7] $\Delta\text{PSNR} = \text{PSNR of fast algorithm} - \text{PSNR of MR-FS algorithm}.$

[8] $\text{Speed-up} = \dfrac{\text{Searched locations for MR-FS algorithm}}{\text{Searched locations for fast algorithm}}.$

Table 8.8: Comparison between different block-matching algorithms in terms of prediction quality (average PSNR$_Y$ in dB) with a multiframe memory of $M = 50$ frames and a frame skip of 1

	AKIYO		FOREMAN		TABLE TENNIS	
	PSNR	ΔPSNR	PSNR	ΔPSNR	PSNR	ΔPSNR
SR-FS	45.93	−0.62	32.20	−1.77	32.17	−0.70
MR-FS	46.55	0.00	33.97	0.00	32.87	0.00
MR-FS/SMS	46.55	0.00	33.92	−0.05	32.80	−0.07
MR-SMS	46.55	0.00	33.87	−0.10	32.67	−0.20
MR-3DSM	46.55	0.00	33.51	−0.46	32.46	−0.41

Table 8.9: Comparison between different block-matching algorithms in terms of computational complexity (average searched locations/frame) with a multiframe memory of size $M = 50$ frames and a frame skip of 1

	AKIYO		FOREMAN		TABLE TENNIS	
	Locations	Speed-up	Locations	Speed-up	Locations	Speed-up
SR-FS	65,621	45.90	77,439	45.90	65,621	45.90
MR-FS	3,012,200	1.00	3,554,700	1.00	3,012,200	1.00
MR-FS/SMS	103,820	29.01	183,240	19.40	134,270	22.43
MR-SMS	38,880	77.47	106,830	33.27	69,443	43.38
MR-3DSM	35,867	83.98	66,357	53.57	45,518	66.18

the MR-3DSM algorithm. Compared to MR-FS, the MR-3DSM algorithm provides significant reductions in computational complexity (a speed-up ratio of about 54–84) at the expense of a moderate reduction in prediction quality (about 0.41–0.46 dB loss[9]). At the other extreme is the MR-FS/SMS algorithm. It uses full search on the most recent reference frame in memory to provide a prediction quality that is almost identical to that of MR-FS (about 0.05–0.07 dB loss) and still achieves moderate reductions in computational complexity (a speed-up ratio of about 22–29). Between the two extremes is the MR-SMS algorithm. Compared to MR-FS, it achieves reasonable reductions in computational complexity (a speed-up ratio of about 33–77) with only a slight loss in prediction quality (about 0.1–0.2 dB loss). These observations are further emphasized using Figure 8.11, which compares the performance of the different algorithms when applied to FOREMAN at different frame skips.

A very interesting point to note (from Tables 8.8 and 8.9 and also from Figure 8.11) is that the computational complexity of the multiple-reference SMS algorithms is comparable to (and in some cases less than) that

---

[9]This excludes the result for AKIYO where ΔPSNR $= 0$.

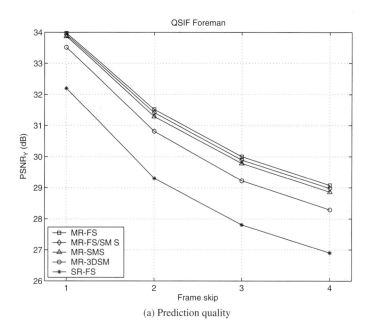

(a) Prediction quality

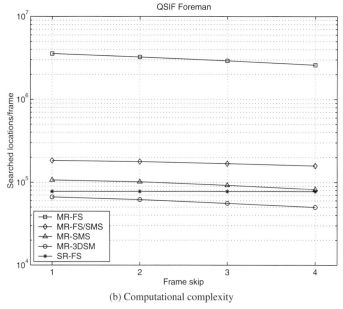

(b) Computational complexity

Figure 8.11: Comparison between different block-matching algorithms when applied to QSIF FOREMAN with a multiframe memory of $M = 50$ frames

(a) Original frame

(b) Compensated using SR-FS (28.24 dB and 77,439 locations)

(c) Compensated using MR-FS with $M=50$ (31.31 dB and 3,871,950 locations)

(d) Compensated using MR-3DSM with $M=50$ (31.04 dB and 72,532 locations)

Figure 8.12: Subjective quality of the motion-compensated 158th frame of QSIF FOREMAN at 25 frames/s

of single-reference full-search SR-FS, and yet they still maintain the improved prediction gain of multiple-reference motion estimation. This is also illustrated in Figure 8.12, which shows the subjective quality of the motion-compensated 158th frame of FOREMAN. The uncovered background at the bottom-right corner of the frame is poorly compensated using the single-reference algorithm SR-FS (Figure 8.12(b)). This uncovered background is compensated with higher quality using the multiple-reference algorithms (Figures 8.12(c) and 8.12(d)). While the MR-FS algorithm achieves this improved prediction quality at the expense of about 50 times increase in computational complexity, the MR-3DSM algorithm provides a similar improvement at no increase in computational complexity.

8.7   Discussion

There are many techniques for reduced-complexity BMME. The most widely used approach employs a reduced set of motion vector candidates. Algorithms in this category are usually based on a *unimodal error surface assumption*. In most cases, however, this assumption does not hold true, and such algorithms can easily get trapped in local minima, giving a suboptimal prediction quality. The main aim of this chapter was to develop a reduced-complexity BMME that adopts the same approach of reducing the set of motion vector candidates but, at the same time, avoids the local minimum problem.

Thus, the chapter formulated block-matching motion estimation as a two-dimensional constrained minimization problem. It was then proposed to solve this problem, with reduced complexity, using an optimization method called the simplex minimization (SM) optimization method. The resulting solution was called the simplex minimization search (SMS). The initialization procedure, termination criterion, and constraints on the independent variables of the search were designed to take into account the basic properties of the BMME problem.

Simulation results within an isolated test environment showed that the SMS algorithm outperforms other reduced-complexity BMME algorithms, providing better prediction quality, smoother motion field, and higher speed-up ratio. In particular, the SMS algorithm is very resilient to the local minimum problem. This superior performance was also confirmed within an H.263-like codec and an object-based MPEG-4 codec.

It was also noted that the superior performance of the LTM-MCP (discussed in Chapter 6) is achieved at the expense of a significant increase in computational complexity. To reduce complexity, the chapter extended the SMS algorithm to the multiple-reference case. Three different extensions (or algorithms) were presented, each representing a different degree of compromise between prediction quality and computational complexity. Simulation results showed that the multiple-reference SMS algorithms provide significant reductions in computational complexity compared to the multiple-reference full-search. With a multiframe memory of size $M = 50$, the computational complexity of the SMS algorithms is comparable to (and in some cases less than) that of single-reference full-search, and yet they still maintain the improved prediction gain of multiple-reference motion estimation.

Part IV
Error Resilience

When transmitted over a mobile channel, compressed video can suffer severe degradation. Thus, error resilience is one of the main requirements for mobile video communication.

This part contains two chapters. Chapter 9 reviews error-resilience video coding techniques. The chapter considers the types of errors that can affect a video bitstream and examines their impact on decoded video. It then describes a number of error detection and error control techniques. Particular emphasis is given to standard error-resilience techniques included in the recent H.263+, H.263++, and MPEG-4 standards.

Chapter 10 gives examples of the development of error-resilience techniques. The chapter presents two temporal error concealment techniques. The first technique, MFI, is based on motion field interpolation, whereas the second technique, BM-MFI, uses multihypothesis motion compensation (MHMC) to combine MFI with a boundary matching (BM) technique. The techniques are then tested within both an isolated test environment and an H.263 codec. The chapter also investigates the performance of different temporal error concealment techniques when incorporated within a multiple-reference video codec. In particular, the chapter finds a combination of techniques, MFI-BM, that best recovers the spatial-temporal components of a damaged multiple-reference motion vector. In addition, the chapter develops a multihypothesis temporal concealment technique, called MFI-MH, to be used with multiple-reference systems.

Chapter 9

Error-Resilience Video Coding Techniques

9.1 Overview

As already discussed, one of the main requirements for mobile video communication is *error resilience*. When transmitted over a mobile channel, video can be affected by a number of loss mechanisms, like multipath fading, shadowing, and co-channel interference. The effects of such errors are magnified due to the fact that the video bitstream is highly compressed to meet the stringent bandwidth limitations. The higher the compression, the more sensitive the bitstream is to errors, since in this case each bit represents a larger amount of decoded video. The effects of errors are also magnified by the use of predictive and VLC coding, which can lead to temporal and spatial error propagation. It is therefore not difficult to realize that when transmitted over a mobile channel, compressed video can suffer severe degradation, making the use of error-resilience techniques vital. This chapter reviews error-resilience video coding techniques.

The rest of the chapter is organized as follows. Section 9.2 describes the main functional blocks of a typical video communication system. Section 9.3 highlights the main types of errors that can affect a video bitstream. Section 9.4 examines the impact of such errors on the decoded video. Section 9.5 describes a number of error detection techniques. Sections 9.6–9.8 reviews three main categories of error-resilience video coding techniques. The chapter concludes with a discussion in Section 9.9.

9.2 A Typical Video Communication System

Figure 9.1 shows a typical video communication system. The *encoder* consists of a *source encoder* and a *channel encoder*.

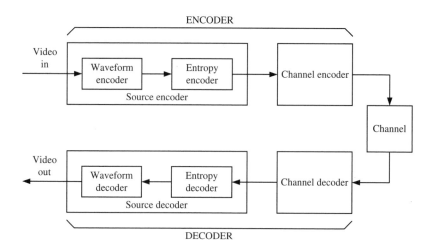

Figure 9.1: Typical video communication system

The function of the source encoder is to compress the input video. It consists of a *waveform encoder* and an *entropy encoder*. The function of the source encoder is described in detail in Chapter 2. With reference to Figure 2.3, the waveform encoder corresponds to the *mapper* and *quantizer* blocks, whereas the entropy encoder corresponds to the *symbol encoder* block. Thus, the waveform encoder works by removing, as much as possible, statistical and psychovisual redundancies present in the input video, whereas the entropy encoder tries to remove coding redundancy.

The channel encoder conditions the compressed bitstream at the output of the source encoder to be suitable for transmission over the channel. This can include, for example, packetization, error protection, modulation, and transport-level control.

At the decoder, the reverse operations are performed to obtain the output video. Note that although this figure shows a one-way communication between the encoder and the decoder, some video communication systems may also have data flowing in the other direction to convey some feedback information.

9.3   Types of Errors

Errors affecting a digital video bitstream can be roughly classified into *two* main categories: *random bit errors* and *erasure errors*.

9.3.1   Random Bit Errors

Random bit errors can occur in the form of bit inversion, bit insertion, and/or bit deletion. They are usually quantified using a parameter called the *bit error rate* (BER), which is the average probability that a bit is in error. Random bit errors are usually caused by physical effects like thermal noise.

9.3.2   Erasure (or Burst) Errors

Erasure errors occur in the form of a loss of (or damage to) contiguous segments of bits. They are usually quantified using parameters like the number of bursts, the length of a burst, and the BER within a burst. Burst errors in a mobile channel can be caused by a number of mechanisms, such as short-term (multipath) fading, long-term (shadowing) fading, and co-channel interference. In a packet-based network, burst errors occur in the form of packet losses due to different reasons, such as congestion, misrouting, and delivery with unacceptably long delays.

It should be pointed out, however, that this classification does not take into account the impact of errors, which is highly dependent on the coding method. For example, it will be shown later that due to the use of predictive and VLC coding, random bit errors in a video bitstream can cause severe error propagation. Thus, random bit errors in a video bitstream are effectively equivalent to burst errors. In what follows, no distinction will be made between the two types of errors, and the generic term *transmission errors* will be used to refer to both types.

9.4   Effects of Errors

Errors occurring in a video bitstream can cause *isolated effects*, *spatial error propagation*, and/or *temporal error propagation*.

9.4.1   Isolated Effects

In this case the effect of an error is limited and does not propagate either spatially or temporally. An example is an error in a FLC codeword. Another example is an error that converts a VLC codeword into another valid codeword of the same length. Note, however, that for both cases to have an isolated effect, it is assumed that the damaged codeword is not a prediction for another codeword and that no temporal error propagation occurs due to motion-compensated prediction. Clearly, such isolated effects are rare occurrences in video bitstreams, and when they do occur their damage is usually acceptable and can be handled relatively easily. However, such errors can sometimes be catastrophic, as, for example, in the case of errors in vital header information (e.g., frame size, and quantizer step size).

9.4.2   Spatial Error Propagation

This is mainly due to two mechanisms:

1. **Errors in VLC Coded Data**: If an error converts a VLC codeword into an invalid codeword or into a valid codeword of a different length, then this causes *loss of bitstream synchronization*. This can occur in two forms [177]:

   (a) *Loss of Codeword Synchronization*: In this case an error causes the decoder to decode a codeword of the wrong length. As a result, the next codeword will be decoded in the wrong position and all following codewords may be affected. This effect is usually temporary, and the decoder eventually regains codeword synchronization [178].

   (b) *Loss of Coefficient Synchronization*: The second form of loss of synchronization is the loss of coefficient synchronization. Even when codeword synchronization is regained, the decoder will be decoding coefficients that have no meaning without the previous, lost coefficients. For example, in run-length encoding, if an incorrect run-length has been decoded, then all the following data will be misplaced even if it is decoded correctly. Since this form of loss of synchronization usually causes data to be misplaced, it is also referred to as *loss of positional synchronization*.

2. **Errors in Predictively Coded Data**: The second mechanism that causes spatial error propagation is the loss of predictively coded data. For example, a motion vector is usually predictively coded with reference to one or more previous motion vectors. If those previous vectors are in error, then the prediction will be wrong and the errors will propagate to the current motion vector, and so on.

9.4.3   Temporal Error Propagation

This is due mainly to the use of motion compensated prediction (or any other form of predictive coding in the temporal dimension). As already described, in motion-compensated prediction, parts of the current frame are copied (or motion compensated) from a reference frame. If the copied reference parts already contain errors, then those errors will also occur in (i.e., propagate to) the current frame.

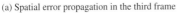

(a) Spatial error propagation in the third frame   (b) Temporal error propagation in the sixth frame

Figure 9.2: Spatial and temporal error propagation due to a single-bit error in QSIF TABLE TENNIS H.263 encoded at 10 frames/s (46 kbits/s)

Figure 9.2 shows an example of spatial and temporal error propagation in the QSIF TABLE TENNIS sequence H.263 encoded[1] at a frame rate of 10 frames/s (about 46 kbits/s). Figure 9.2(a) shows the third frame of the sequence, where a single bit error hits the macroblock in the position shown. This error converts the VLC codeword representing the vertical vector difference to another valid codeword of the same length. This causes an error in the compensation of this particular macroblock. In addition, because of the predictive coding of motion vectors, this error propagates spatially to all macroblocks to the right and up to the border of the frame. Figure 9.2(b) shows how motion-compensated prediction caused the errors in the third frame to propagate temporally to the sixth frame. This example shows how serious even a single bit error can be and clearly highlights the need for error detection and control techniques.

9.5 Error Detection

Before being able to combat the effects of errors, it is first necessary to detect whether and where errors have occurred. Error detection can be performed by the channel decoder and/or the source decoder.

One method for error detection is the use of header information. This can be used by both the channel decoder and the source decoder. For example, in a packet-based network like ATM, each packet contains a header with a

---

[1]Telenor H.263 implementation was used. The luma component was zero padded to 128 lines to be a multiple of 16. The chroma components were also zero padded correspondingly. The optional mode to insert synchronization codewords at the start of each GOB was switched on. All other optional modes were switched off. The initial quantization parameter was set to 10.

sequence number subfield. This sequence number can be used to detect packet losses at the channel decoder. Similarly, the group-number (GN) codeword in an H.263 GOB header can be used to detect errors at the source decoder.

Another method that can be used by both the channel decoder and the source decoder is *forward error correction* (FEC). For example, Annex H of the H.263 standard provides an optional FEC mode. In this mode, 18 parity bits are used to provide error detection and correction for each 493 video bits.

A commonly used method at the source decoder is the detection of syntax and semantic violations. Examples of such violations are:

- An illegal codeword is detected.

- An invalid number of units is decoded. For example, the number of decoded DCT coefficients within a block is invalid, the number of decoded blocks within a MB is invalid, the number of decoded MBs within a GOB is invalid, or the number of decoded GOBs within a frame is invalid.

- A decoded motion vector points outside the permissible range.

- A decoded quantization parameter is out of range.

Another method that can be used at the source decoder is the detection of violations to the general characteristics of natural video signals, for example, the detection of strong discontinuities at the borders of blocks, blocks with highly saturated colours (e.g., pink and green), or blocks where most pels need clipping.

None of these methods guarantee finding all errors within a video bitstream. In fact, the last method may sometimes detect an error-free block as an erroneous one. In practical systems, different combinations of these methods are employed.

Having detected the occurrence of errors and identified their locations, a number of methods can be used to combat the effects of errors on the video bitstream. The following three sections describe three categories of error-resilience techniques: *forward* techniques, *postprocessing* techniques, and *interactive* techniques. The three sections follow closely the classification used in the comprehensive reviews by Wang *et al.* [179, 180].

9.6   Forward Techniques

In forward techniques, the *encoder* plays the primary role. Such techniques work by adding a controlled amount of redundancy to the video bitstream. This means that they sacrifice some coding efficiency to gain in terms of error resilience. Some techniques are designed to minimize the effects of transmission errors, some are designed to make error handling at the decoder more effective, and others are designed to guarantee a basic level of quality while providing graceful degradation in the presence of transmission errors. Examples of forward techniques are briefly described in the following subsections.

9.6.1 Forward Error Correction (FEC)

Forward error correction works by adding redundant bits to a bitstream to help the decoder detect and correct some transmission errors without the need for retransmission. The name *forward* stems from the fact that the flow of data is always in the forward direction (i.e., from encoder to decoder).

For example, in block codes the transmitted bitstream is divided into blocks of k bits. Each block is then appended with r parity bits to form an n-bit codeword. This is called an (n, k) code.

For example, Annex H of the H.263 standard provides an optional FEC mode. This mode uses a $(511, 493)$ BCH (*Bose-Chaudhuri-Hocquenghem*) code. Blocks of $k = 493$ bits (consisting of 492 video bits and 1 fill indicator bit) are appended with $r = 18$ parity bits to form a codeword of $n = 511$ bits. Use of this mode allows the detection of double-bit errors and the correction of single-bit errors within each block.

9.6.2 Robust Waveform Coding

As already discussed, the waveform encoder in a typical video communication system works by removing statistical and psychovisual redundancies present in the input video. Robust waveform coding techniques, however, intentionally keep (or even add) some redundancy to achieve error resilience. Examples of such techniques are given next.

9.6.2.1 Adding Redundant Information

This technique adds auxiliary information or repeats some previously coded information to help error handling at the decoder. For example, as is shown in Section 9.7, a powerful technique for error concealment is temporal concealment. The performance of this technique is highly dependent on the availability of motion information for the damaged blocks. Thus, this technique is usually used for concealing INTER macroblocks. In MPEG-2, however, the encoder can optionally send auxiliary motion vectors for INTRA macroblocks. In the presence of errors, such vectors can be used to temporally conceal damaged macroblocks.

Another example is the header extension code (HEC) included by MPEG-4 in packet headers. If this bit is set to "1," then some data, like timing information and VOP coding type, is repeated from the VOP header. This helps error detection and resynchronization. A similar example is the picture header repetition allowed by the optional additional supplemental enhancement information mode (annex W) of H.263++.

9.6.2.2   Using INTRA Refresh

An effective way to stop temporal error propagation is to periodically encode pictures in INTRA mode. Given the large number of bits consumed by IN-TRA pictures, this leads to a significant increase in the total bit rate. A more suitable approach for applications like mobile video communication is to use INTRA refresh on the macroblock level. By controlling the *number* and *spatial location* of INTRA MBs, INTRA refresh can be a very efficient and scalable error-resilience tool.

Obviously, the required number of INTRA MBs is highly dependent on the channel quality and capacity. Such information is usually available to the encoder. For example, in mobile networks, antenna parameters can give an indication of the channel quality. In Ref. 181, Haskell and Messerschmitt discuss how to select a suitable number of INTRA MBs.

There are many methods for selecting the spatial location of INTRA MBs within frames. One method is to choose the locations randomly [181, 182]. Another method is to follow a raster scanning order. In Ref. 183 the INTRA MBs are placed adaptively in regions with high activity.

Recently, a very powerful technique for deciding both the number and spatial locations of INTRA MBs has been proposed by Côté *et al.* [182, 184]. In Ref. 182 they propose a rate-distortion optimized mode selection method for packet lossy networks. This method takes into account the channel conditions and the error concealment method used at the decoder. In Ref. 184 they apply the same method to bit-oriented networks.

Obviously, if there is a feedback channel from the decoder, then information regarding the number and locations of damaged MBs can help the encoder to better decide the number and locations of INTRA MBs.

9.6.2.3   Using Restricted Prediction

In this technique, prediction is limited within nonoverlapping spatial and/or temporal regions. This clearly limits temporal and/or spatial error propagation.

For example, in the independent segment decoding mode (annex R) of H.263+, video pictures are divided into segments. Each video picture segment is then encoded with complete independence from all other segments in the same picture, and also with complete independence from all data outside the corresponding segment in the reference picture(s). For example, motion vectors of blocks outside the current segment cannot be used when calculating the current motion vector predictor. Similarly, motion vectors of blocks outside the current segment cannot be used as remote motion vectors for overlapped block-motion compensation when the advanced prediction mode is in use. In addition, no motion vectors are allowed to reference areas outside the corresponding segment in the reference picture.

9.6.3 Robust Entropy Coding

In this case, redundancy is added at the entropy encoder. Examples of robust entropy coding techniques are discussed next.

9.6.3.1 Resynchronization Codewords

As already discussed, one of the disadvantages of VLC coding is that errors in the bitstream can cause loss of synchronization between the encoder and the decoder, and this leads to spatial error propagation. One way to reduce this effect is to insert unique markers called *resynchronization codewords* in the bitstream. When an error is detected, the decoder skips the remaining bits until it finds a resynchronization codeword. This reestablishes synchronization with the encoder, and the decoder then proceeds to decode from that point on. This is illustrated in Figure 9.3(a).

Resynchronization codewords can be inserted at regular intervals in the spatial domain, as illustrated in Figure 9.4(a). For example, version 1 of H.263 adopts a GOB-based resynchronization approach. This means that a resynchronization codeword is inserted every time a fixed number of macroblocks has been encoded. A disadvantage of this approach is that, since the number of bits can vary between macroblocks, the resynchronization codewords will most likely be unevenly spaced throughout the bitstream. Therefore, certain parts of the sequence, such as high-motion areas with high bit content, will be more susceptible to errors and will also be more difficult to conceal.

A more robust approach is to insert resynchronization codewords at regular intervals in the bit domain, as illustrated in Figure 9.4(b). For example, MPEG-4 adopts a packet-based resynchronization approach. In this approach each packet contains approximately the same number of bits. This means that the resynchronization codewords are almost periodic in the bitstream. A similar approach has also been adopted in the slice structured mode (annex K) of H.263+.

Another problem with VLC coding is that errors can emulate the occurrence of resynchronization codewords. To reduce this effect, MPEG-4 provides a second resynchronization approach called *fixed-interval synchronization*. In this approach, resynchronization codewords appear only at legal fixed-interval

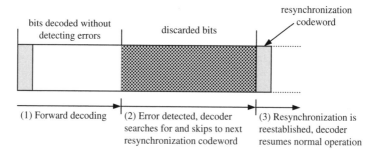

(a) Resynchronization with normal VLC coding

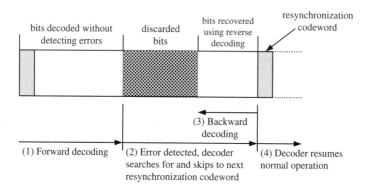

(b) Resynchronization with reversible VLC (RVLC) coding

Figure 9.3: Resynchronization using synchronization codewords

locations in the bitstream. Thus, only codewords at those legal locations will be used by the decoder to reestablish synchronization.

As described in Section 9.4, loss of synchronization appears in two forms: loss of codeword synchronization and loss of positional (or coefficient) synchronization. Inserting resynchronization codewords reduces the effect of loss of codeword synchronization. In order to reduce the effect of loss of positional synchronization, resynchronization codewords are usually followed by some positional information, like the address and the temporal reference of the macroblock immediately following the resynchronization codeword. This allows the decoder to resume its normal operation.

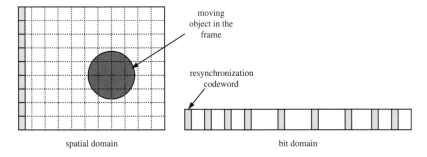

spatial domain   bit domain

(a) Resynchronization codewords at regular intervals in the spatial-domain

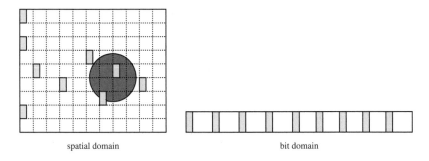

spatial domain   bit domain

(b) Resynchronization codewords at regular intervals in the bit-domain

Figure 9.4: Resynchronisation codewords at regular intervals

9.6.3.2   The Error-Resilience Entropy Code (EREC)

An interesting alternative to inserting resynchronisation codewords is the *error resilience entropy code* (EREC) [177, 185]. The EREC takes variable-length blocks of data and rearranges them into fixed-length slots. For example, assume that there are N variable-length blocks with lengths b_i, $i = 1 \ldots N$. The encoder first chooses a total data size $T \geqslant \sum b_i$, which is sufficient to encode all the data. This total data size is split into N slots of fixed lengths s_i, $i = 1 \ldots N$. An N-stage algorithm is then used to place the data from the variable-length blocks into the fixed-length slots. At each stage n, a block i with data left unplaced searches slot $j = i + \phi_n \pmod{N}$ for space to place some or all of the remaining data. Here, ϕ_n is an offset sequence that is usually pseudo-random.

Figure 9.5 shows an example of the EREC algorithm. In this case, there are $N = 6$ variable-length blocks, with lengths 11, 9, 4, 3, 9, and 6 bits. The total data size is chosen as $T = 42$ and is divided into $N = 6$ slots, with a length of $s_i = 7$ bits each. The offset sequence is $\phi_n = \{0, 1, 2, 3, 4, 5, 6\}$. In stage 1 of

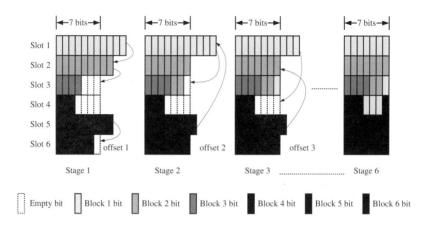

Figure 9.5: Example of the EREC algorithm

the algorithm, blocks 3, 4, and 6 are completely placed into the corresponding slots, with some leftover space in those slots. Blocks 1, 2, and 5, however, are only partially placed in the corresponding slots and have some bits left to be placed in empty spaces in other slots. According to the offset sequence, block 1 searches slot 2 for empty space, block 2 searches slot 3, and block 5 searches slot 6. Both blocks 2 and 5 find empty spaces. Thus, in stage 2, all the remaining bits from block 2 are placed in slot 3, whereas some of the remaining bits of block 5 are placed in slot 6. Since block 1 did not find empty spaces in slot 2, then, according to the offset sequence, it searches slot 3, and so on. By the end of stage 6, all data bits are placed in the slots. The decoder operates in a similar manner. Thus bits in a slot are decoded and placed in a block until an end-of-block codeword is encountered.

In the presence of errors, the resilience provided by the EREC algorithm is due to two factors. First, each block starts at a known position in the bitstream (i.e., the start of the corresponding slot). Thus, in the case of loss of synchronization, the decoder simply jumps to the start of the next slot without the need for resynchronization codewords. Second, subjectively less important data (e.g., high-frequency DCT coefficients) are usually placed in later stages of the algorithm. With the EREC algorithm, most error propagation effects (due, for example, to missing or falsely detecting end-of-block codewords) hit data placed at later stages of the algorithm rather than the more important data at the start of the slots.

9.6.3.3 Reversible Variable-Length Coding (RVLC)

Reversible VLC codewords are designed to be decoded both in the forward and backward directions. As already described, when an error is detected in the bitstream, the decoder discards all bits until the next resynchronization codeword, where synchronization is reestablished and the decoder resumes its decoding process. The discarded bits may well be correctly received but cannot be decoded correctly due to loss of synchronization. In the case of RVLCs, when the decoder identifies the next resynchronization codeword, instead of discarding all preceding bits, the decoder starts decoding in the reverse direction to recover and utilize some of those bits. This is illustrated in Figure 9.3(b).

Reversible variable-length coding has been adopted in most recent standardization efforts. For example, the modified unrestricted motion vector mode (modified annex D) of H.263+ uses RVLC to encode motion vector differences, the data partitioned slice mode (annex V) of H.263++ uses RVLC to encode header and motion information, and MPEG-4 uses RVLC to encode texture information.

9.6.4   Layered Coding with Prioritization

In layered coding, video is encoded into a base layer and one or more enhancement layers. The base layer is separately decodable and provides a basic level of perceived quality. The enhancement layers can be decoded to incrementally improve this quality.

Layered coding can be useful when applied over heterogenous networks with varying bandwidth capacity. However, to be used as an error-resilience tool, layered coding must be combined with prioritized transmission or what is commonly known as *unequal error protection*. In this case, the base layer is transmitted with higher priority or a higher degree of error protection. For example, in Ref. 186 Ghanbari introduced the concept of layered coding with prioritized transmission to increase the robustness of video against cell loss in ATM networks. In this technique, the encoder generates two bitstreams. The base-layer bitstream contains the most vital video information, whereas the enhancement-layer bitstream contains residual information to improve the quality of the base layer. The base layer is then transmitted using high-priority ATM cells, whereas the enhancement layer is transmitted using low-priority cells. When traffic congestion occurs, low-priority cells are discarded first. Another example is the power control method proposed in Ref. 187. In this method, when video is transmitted over a wireless network, more power is used to transmit the base layer, whereas less power is used to transmit the enhancement layers.

There are many ways to encode video into more than one layer. For example, the base layer can include a low-frame-rate version of video, whereas the enhancement layers can contain frames used to increase the frame rate. This is usually referred to as *temporal scalability*. Another method is when the base layer contains a coarsely quantized version of video, whereas the enhancement layers carry the error between the original version and this coarsely quantized version. This is known as *SNR scalability*. Another form is *spatial scalability*. This is very similar to SNR scalability. The only difference is that pictures in the base layer are subsampled to a smaller size. Yet another form of layered coding is known as *data partitioning*. In this case, the base layer contains vital video information like headers, motion vectors, and low-frequency DCT coefficients. Other information, like high-frequency DCT coefficients, is included in the enhancement layers.

Note that all these forms of layered coding are supported in recent standardization efforts. For example, MPEG-4 supports temporal and spatial scalability in addition to data partitioning. H.263+ supports temporal, SNR, and spatial scalability in annex O, and H.263++ supports data partitioning in annex V.

9.6.5 Multiple Description Coding

This technique assumes that there are multiple channels between the encoder and the decoder. These multiple channels can be physically distinct paths or they can be a single path divided into multiple virtual channels using, for example, time or frequency division. The technique further assumes that the error events of these multiple channels are independent. This means that the probability that all channels simultaneously experience errors is very small.

Similar to layered coding, multiple description coding encodes video into multiple streams known as *descriptions*. In this case, however, the descriptions are correlated and have equal importance. The requirement that all descriptions have equal importance means that the descriptions must share some fundamental information about the input video. As a consequence of this information sharing, the descriptions are correlated.

At the encoder, each description is transmitted on a different channel. As already mentioned, the error events of the channels are independent. As a result, at least one description will be received at the decoder without errors. This description carries some fundamental information about the transmitted video and can, therefore, be used to provide a basic level of quality. Since the descriptions are correlated, missing descriptions can be estimated from correctly received descriptions and the quality can be improved.

There are a number of methods to achieve the required decomposition into descriptions. For example, in Ref. 188, the input signal is decomposed and encoded into two streams. The two streams are obtained by transmitting two quantization indices for each quantized level. The index assignment is designed such that when both indices are received, the reconstruction quality is equivalent to that of a fine quantizer. When, however, only one index is received, the reconstruction quality is equivalent to that of a coarse quantizer.

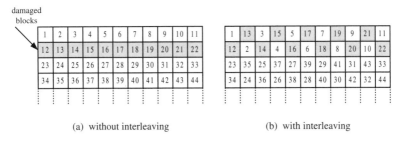

(a) without interleaving   (b) with interleaving

Figure 9.6: Coding and transmission order with and without interleaving

There are also other multiple description techniques, as detailed in Refs. 179 and 180.

9.6.6 Interleaved Coding

In normal coding, the blocks of a given frame are encoded in raster scan order, as illustrated in Figure 9.6(a). In this case, when an error occurs in one block, spatial error propagation results in the loss of a contiguous set of blocks. In the example shown, an error in block 12 results in the loss of all blocks to its right.[2] As is discussed later, the concealment of a damaged block depends heavily on the availability of its four neighboring blocks. In this case, a damaged block will have only its top and bottom neighbors intact.

Interleaved coding attempts to separate the information of neighboring blocks as far as possible. As a result, an error in a block will propagate to nonadjacent blocks. Figure 9.6(b) shows the even/odd interleaving scheme adopted in Ref. 189. The numbers here indicate the encoding and transmission order. Thus, the first block in the first row (block 1) is encoded and transmitted first, followed by the second block in the second row (block 2), and so on. Note that in this case, when an error occurs in block 12, the lost set of blocks is not contiguous. Thus, a damaged block will have all its four neighbors intact and this will help the error concealment process considerably.

9.7 Postprocessing (or Concealment) Techniques

The second category of error-resilience techniques are postprocessing (or concealment) techniques. In postprocessing techniques, the *decoder* plays the

---

[2]This example assumes that resynchronization codewords are inserted at the beginning of each row of blocks.

primary role. Thus, the decoder attempts to *conceal* the effects of errors by providing a *subjectively* acceptable approximation to the original data. This is achieved by exploiting the limitations of the human visual system and the high temporal and/or spatial correlation of video sequences. Error concealment is an ill-posed problem since it does not have a unique solution. Thus, error concealment techniques exploit *a priori* knowledge of the characteristics of video signals to restrict the otherwise large number of possible solutions. Depending on the information used for concealment, postprocessing techniques can be divided into three main categories: *spatial* techniques, *temporal* techniques, and *hybrid* techniques.

9.7.1 Spatial Error Concealment

Spatial techniques exploit the high spatial correlation of video signals and conceal damaged pels in a frame using information from correctly received and/or previously concealed neighboring pels within the same frame. Such techniques apply primarily to intracoded blocks but may also be used to conceal intercoded blocks with missing motion information or to recover the DFD signal.

In Ref. 190 a damaged pel within a block is interpolated from the four corner pels outside the block, as illustrated in Figure 9.7(a). Interpolation from the four nearest pels outside the block boundaries, as illustrated in Figure 9.7(b), is proposed in Ref. 191. Interpolation in the frequency domain has also been used. For example, in Ref. 192 the DC coefficient of a damaged block is recovered as the average or the median of the DC

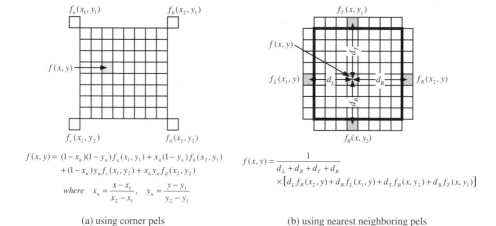

$$f(x,y) = (1-x_n)(1-y_n)f_a(x_1,y_1) + x_n(1-y_n)f_b(x_2,y_1)$$
$$+ (1-x_n)y_nf_c(x_1,y_2) + x_ny_nf_d(x_2,y_2)$$
$$\text{where} \quad x_n = \frac{x-x_1}{x_2-x_1}, \quad y_n = \frac{y-y_1}{y_2-y_1}$$

$$f(x,y) = \frac{1}{d_L + d_R + d_T + d_B}$$
$$\times [d_Lf_R(x_2,y) + d_Rf_L(x_1,y) + d_Tf_B(x,y_2) + d_Bf_T(x,y_1)]$$

(a) using corner pels   (b) using nearest neighboring pels

Figure 9.7: Error concealment using spatial interpolation

coefficients of the four or eight neighboring blocks. Another approach is to form a partial DC value at each boundary by taking the average of a one-, two-, or four-pels-wide neighborhood. The recovered DC coefficient is then the average or the median of the four partial DC values.

In Ref. 193 the lost DCT coefficients of an intracoded block are recovered by minimizing the intersample variation within the block and across the block boundaries. This is based on the smoothness property of image and video sequences. In Ref. 189 the same method is extended by adding a temporal smoothness measure.

Another property that is used in error concealment is edge continuity. Thus, if the direction of an edge in a neighboring block indicates that the edge passes through the damaged block, then the concealment process must conserve the continuity of this edge. For example, in Ref. 194 an edge classifier is applied to the neighboring blocks to determine which directions characterize the strongest edges passing through the damaged block. For each of these classified directions, directional spatial interpolation along the respective direction is used to create a block from the neighboring pels. The blocks are then mixed together in such a way that all the strong edge features are preserved and combined in a single block used for concealment.

Statistical correlation is another *a priori* assumption utilized in error concealment. For example, in Ref. 195 the pel values of a frame are modeled as a Markov random field (MRF). Maximum *a posteriori* probability (MAP) estimation is then used to spatially interpolate the damaged blocks.

9.7.2   Temporal Error Concealment

Temporal techniques exploit the high temporal correlation of video signals and conceal damaged pels in a frame using information from correctly received and/or previously concealed pels within a reference frame. Such techniques apply primarily to intercoded blocks. They may work for some intracoded blocks but will completely fail in cases like scene changes and uncovered background.

As in motion-compensated prediction, the process of temporal concealment involves two stages: *concealment displacement estimation* and *displacement compensation*, as shown in Figure 9.8(a). For this reason, temporal concealment is sometimes referred to as *motion-compensated concealment*.

Conventional temporal techniques estimate one concealment displacement for the whole damaged block and then use translational displacement compensation to conceal the block, as shown in Figure 9.8(b). Such techniques perform very well when the original motion vector of the damaged block is available. In this case the first stage of the temporal concealment process,

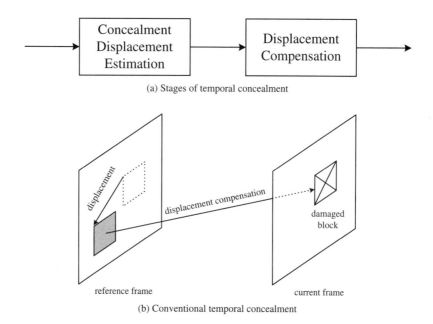

(a) Stages of temporal concealment

(b) Conventional temporal concealment

Figure 9.8: Temporal error concealment

i.e., displacement estimation, is bypassed and the concealment displacement is simply set to the original motion vector.

In practice, however, the motion vector of a damaged block is usually lost or erroneously received. This is due mainly to spatial error propagation. For example, an erroneous codeword will usually lead to loss of synchronization at the decoder and all blocks, including their motion information, up to the next synchronization point will be undecodable and completely lost.[3] In such cases, the displacement estimation stage at the decoder is extremely important. In fact, the only difference between the various conventional temporal techniques reported in the literature is in their displacement estimation algorithm. This stage is also known as *motion information recovery*, because it attempts to recover or provide an approximation to the original motion information.

The simplest and most commonly used technique is to replace the damaged motion vector with $(0,0)$ [179, 192]. This is based on the center-biased property of video block-motion fields, which is also equivalent to the temporal smoothness property of video signals. The technique is usually referred to as

---

[3]As already discussed, RVLCs and data partitioning into motion and texture data are some of the mechanisms that can be used to reduce this effect.

temporal replacement (TR) because it effectively replaces the damaged block by its corresponding block in the reference frame. This method works well for stationary and quasi-stationary areas, e.g., background, but will fail for fast-moving areas.

Another technique is to exploit the high-correlation property of video block-motion fields and replace the damaged motion vector with the *average* (AV) [179, 190, 189, 191, 192] or the median [179, 192] of neighboring vectors. This technique works well for areas with smooth motion but will fail for areas with unsmooth motion, e.g., at the boundaries of objects moving in different directions.

A boundary matching (BM) technique has also been used to select a suitable replacement from a set of candidate motion vectors [196, 197, 198]. Assume that a set of M neighboring motion vectors $\mathscr{V} = \{\mathbf{v}_1, \mathbf{v}_2, \ldots, \mathbf{v}_M\}$ is to be used for the concealment of a damaged block $\mathscr{D}$ of size $N \times N$ with its top-left corner at (x_o, y_o). Each candidate vector $\mathbf{v}_i = (v_i^x, v_i^y)$ in $\mathscr{V}$ is used to conceal the damaged block $\mathscr{D}$. The quality of this concealment is assessed using the continuity across the concealed block boundaries. This continuity is measured using the *side-match distortion* (SMD) measure, defined as

$$\text{SMD}_i = \text{SMD}_i^L + \text{SMD}_i^R + \text{SMD}_i^T + \text{SMD}_i^B, \tag{9.1}$$

where SMD_i^L is the sum of absolute, or squared, differences across the *left* boundary of block $\mathscr{D}$ when concealed using candidate vector $\mathbf{v}_i$. Thus

$$\text{SMD}_i^L = \sum_{k=0}^{N-1} g[f_t(x_o - 1, y_o + k) - f_{t-\Delta t}(x_o + v_i^x, y_o + v_i^y + k)], \tag{9.2}$$

where f_t and $f_{t-\Delta t}$ are the current and reference frames, respectively, $g = (\cdot)^2$ for the SSD, and $g = |\cdot|$ for the SAD. Similarly, SMD_i^R, SMD_i^T and SMD_i^B are the side-match distortions across the *right*, *top*, and *bottom* boundaries, respectively. Based on the smoothness property of video signals, the candidate motion vector that achieves the minimum SMD is chosen as the recovered motion vector. Thus

$$\hat{\mathbf{v}} = \arg \min_{\mathbf{v}_i \in \mathscr{V}} \text{SMD}_i. \tag{9.3}$$

The main advantage of this method is that displacement estimation is based on a distortion measure. The method will fail for areas with unsmooth motion and also for areas with low spatial correlation, e.g., at the boundaries of objects.

Similar to spatial concealment, Bayesian statistical approaches have also been used for motion vector recovery, e.g., Ref. 195.

9.7.3   Hybrid Error Concealment

Hybrid techniques exploit both spatial and temporal correlations of video signals. A straightforward technique is to use spatial concealment for intracoded blocks and temporal concealment for intercoded blocks. More sophisticated combinations are also possible. For example, in Ref. 199 temporal concealment is first used to get an initial estimate of the damaged block. This initial estimate is then refined using spatial concealment.

9.7.4   Coding-Mode Recovery

As already discussed, each of the preceding concealment techniques applies to a particular type of macroblocks. More specifically, spatial concealment is more applicable to intracoded blocks, whereas temporal concealment is more suitable for intercoded blocks. Provided that the coding mode of a damaged block is known, the appropriate type of concealment is applied. In many cases, however, the coding-mode information of a damaged block is also damaged. Thus, coding-mode information needs to be recovered first before being able to choose the appropriate concealment method.

In Ref. 189, when the coding mode is damaged it is simply set to INTRA and the corresponding block is concealed using spatial techniques.

Usually, there is a high correlation between the coding modes of adjacent blocks. Thus, the coding mode of a damaged block can be estimated from the coding modes of neighboring blocks. In Ref. 200, the coding mode of a damaged MB in an MPEG-2 coded video is estimated from the coding modes of its top and bottom neighboring MBs. For example, the coding mode of a damaged MB in a P-frame is set to INTRA only if its top and bottom neighboring MBs are both INTRA coded; otherwise, a FORWARD INTER mode is assumed.

9.8   Interactive Techniques

The third type of error-resilience methods are interactive techniques. In this case, the *encoder* and *decoder* cooperate to minimize the effects of transmission errors. In such techniques, the decoder uses a feedback channel to inform the encoder about which parts of the transmitted video have been received in error. Based on this feedback information, the encoder adjusts its operation to combat the effects of such errors. The following subsections discuss some examples of interactive (or feedback-based) techniques. A more comprehensive review of such techniques can be found in Ref. 201.

9.8.1 Automatic Repeat Request (ARQ)

In this technique, when an error is detected, the decoder automatically requests the encoder to retransmit the damaged data. When this ARQ is received, the encoder retransmits the requested data. Usually, this retransmission is repeated until either the requested data is correctly received or a predetermined number of retransmissions is exceeded.

Typically, when a decoder sends an ARQ, it waits for the arrival of the requested data before resuming normal operation. This introduces delays that may not be acceptable in real-time applications like mobile video communication. To overcome such delays, Wang and Zhu [179] proposed a technique called *retransmission without waiting*. In this technique, instead of waiting for the arrival of the requested data, the damaged video part is concealed and normal decoding operation is then resumed. A trace of the affected pels and their associated coding information is recorded until the arrival of the requested data. This error trace, along with the received data, is then used to correct the affected pels. Another technique proposed in Ref. 179 is the *multicopy retransmission*. In this technique, multiple copies of the damaged data are sent in each single retransmission trial. This reduces the required number of retransmissions and, consequently, reduces delays.

9.8.2 Error Tracking

When feedback information is received, the encoder can reconstruct the error propagation process. In other words, the encoder can *track* the error propagation from the original occurrence up to the current frame. A number of techniques can then be used to utilize this error trace, as discussed next.

9.8.2.1 INTRA Refresh Based on Feedback

Based on the error trace, areas in the current frame that would have been predicted from affected pels in the reference frame are INTRA encoded. This is illustrated in Figure 9.9. Figure 9.9(a) shows the spatial and temporal propagation in a sequence of frames due to an error in frame n. In Figure 9.9(b) a feedback message arrives at the encoder before the time to encode frame $n + d$. The encoder tracks this error and the affected pels from frame n up to frame $n + d - 1$. During the encoding process of the current frame, $n + d$, blocks that would have been predicted from affected pels in the reference frame, $n + d - 1$, are encoded in INTRA mode to stop error propagation to the next frame, $n + d + 1$.

There are two main drawbacks to this approach. First, a perfect reconstruction of error propagation is a computationally complex process. Second, in cases of high error rates, INTRA refresh can result in a significant loss in

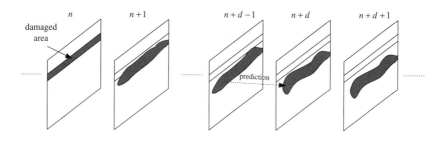

(a) Error propagation in a sequence of frames

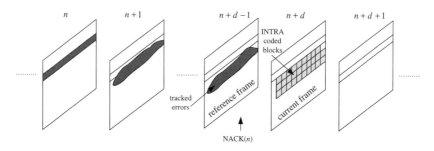

(b) INTRA refresh based on error tracking and feedback information

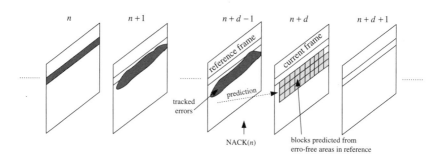

(c) Restricted prediction based on error tracking and feedback information

Figure 9.9: Error tracking techniques

coding efficiency. In Ref. 202 Steinbach *et al.* propose a reduced-complexity error-tracking algorithm that can provide a sufficiently accurate estimate of the true error propagation. In order to reduce the loss of coding efficiency, they INTRA refresh only severely affected blocks. Thus, if the process of error concealment is successful and the error of a given block is sufficiently small, then the encoder may decide against INTRA encoding. Note that this method requires the encoder to perform the same error concealment process that was used at the decoder.

9.8.2.2   Restricted Prediction Based on Feedback

Based on the error trace, prediction of the current frame is restricted to use only error-free areas in the reference frame. For example, in Figure 9.9(c) the affected pels in the reference frame, $n + d - 1$, are not used for predicting the current frame, $n + d$. This stops error propagation to the next frame, $n + d + 1$. This restricted prediction based on feedback and error tracking was proposed by Wada in the *selective recovery* technique [203].

Again, this technique can also benefit from the reduced-complexity error-tracking algorithm of Steinbach *et al.* [202], and the coding efficiency can also be improved by performing error concealment in the encoder so that both encoder and decoder use the same reference frames for prediction.

9.8.3   Reference Picture Selection

In reference picture selection (RPS), both the encoder and decoder store multiple previous frames to be used as reference frames. When the encoder learns, through feedback messages from the decoder, that the most recent reference frame contains errors, the encoder switches to use another older reference frame that is known to be error free. Provided the alternative reference frame is not too far away from the current frame, the loss in coding efficiency is not significant. In particular, this technique is more efficient than the INTRA refresh technique. The RPS technique has been adopted by H.263+ in annex N, and an enhanced version of the technique has been included in annex U of H.263++.

Figure 9.10 shows the RPS technique with two types of feedback messages. In the negative acknowledgment mode, illustrated in Figure 9.10(a), the decoder sends a *negative acknowledgment* (NACK) message whenever errors are detected in a frame. In the example shown, the decoder detects an error in frame 3 and sends a NACK(3) message to the encoder. At the encoder, the encoding operation proceeds in the normal way (i.e., using the most recent reference frame for prediction) until the NACK(3) message arrives before encoding frame 6. Based on this message, the encoder knows that errors occurred in frame 3 and propagated up to the most recent reference frame 5. To stop this error propagation, the encoder uses the older error-free reference frame 2 instead of the most recent reference frame 5 to encode the current frame 6.

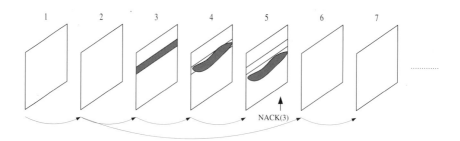

(a) Reference picture selection with negative acknowledgment messages

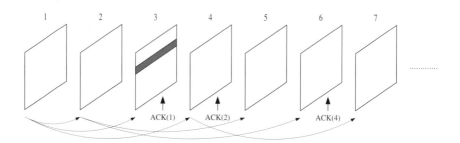

(b) Reference picture selection with positive acknowledgment messages

Figure 9.10: Reference picture selection based on feedback

In the positive acknowledgment mode, illustrated in Figure 9.10(b), the decoder sends an *acknowledgment* (ACK) message whenever a frame is received error-free. At the encoder, only acknowledged frames are used as references. In the example shown, the encoder continues to use frame 1 for prediction until it receives the acknowledgment for frame 2. The encoder then starts using the acknowledged frame 2 for prediction until the acknowledgment of the next error-free reference frame is received. Note that since the erroneous frame 3 is not acknowledged, it is never used for prediction and its errors do not propagate to subsequent frames.

Note that during error-free transmission, the NACK mode is more efficient than the ACK mode since the most recent reference frame is used for prediction. During erroneous transmission, however, the NACK mode results in longer periods of error propagation than the ACK mode. Thus, the NACK mode is more suitable if errors occur only rarely after long periods of error-free transmission, whereas the ACK mode is preferred for highly error-prone transmissions.

9.9 Discussion

When transmitted over a mobile channel, compressed video can suffer severe degradation. Thus, error resilience is one of the main requirements for mobile video communication.

Due to the use of predictive and VLC coding, transmission (both random and erasure) errors cause temporal and spatial error propagation in compressed video.

Before being able to combat these effects, it is first necessary to detect whether and where errors have occurred. Different techniques can be used to achieve this error detection.

Error control techniques can be broadly classified into three categories: forward, postprocessing, and interactive techniques. In forward techniques, the encoder plays the primary role. Such techniques work by adding a controlled amount of redundancy to the video bitstream. In postprocessing techniques, the decoder plays the primary role. Thus, the decoder attempts to conceal the effects of errors by providing a subjectively acceptable approximation to the original data. This is achieved by exploiting the limitations of the human visual system and the high temporal and/or spatial correlation of video sequences. In interactive techniques, the encoder and decoder cooperate to minimize the effects of transmission errors. In such techniques, the decoder uses a feedback channel to inform the encoder about which parts of the transmitted video have been received in error. Based on this feedback information, the encoder adjusts its operation to combat the effects of such errors.

It should be emphasized that the three categories of techniques are not mutually exclusive, and different combinations can be employed in practical systems.

Chapter 10

Error Concealment Using Motion Field Interpolation

10.1 Overview

Chapter 9 discussed three categories of error-resilience techniques: forward, postprocessing (or concealment), and interactive techniques. Almost all forward techniques increase the bit rate because they work by adding redundancy to the data, e.g., FEC. Some of them may also require modifications to the encoder, e.g., layered coding, and others may not be suitable for some applications, e.g., multiple description coding assumes several parallel channels between transmitter and receiver. Most interactive techniques depend on a feedback channel between the encoder and decoder. Such a channel may not be available in some applications, e.g., multipoint broadcasting. Most interactive techniques will also introduce some delay and may, therefore, be unsuitable for real-time applications like mobile video communication. On the other hand, concealment techniques do not increase the bit rate, do not require any modifications to the encoder, do not introduce any delay, and can be applied in almost any application. This makes them a very attractive choice for mobile video communication, where bit rate and delay are very critical issues.

A very successful class of error concealment is *temporal error concealment*. Conventional temporal concealment techniques estimate one concealment displacement for the whole damaged block and then use translational displacement compensation to conceal the block from a reference frame. The main problem with such techniques is that incorrect estimation of the concealment displacement can lead to poor concealment of the whole or most of the block.

This chapter describes the design of two novel temporal concealment techniques. In the first technique, *motion field interpolation* (MFI) is used to estimate one concealment displacement per pel of the damaged block. Each

231

pel is then concealed individually. In this case, incorrect estimation of a concealment displacement will affect only the corresponding pel. On a block level, this may affect few pels rather than the entire block. In the second technique, *multihypothesis motion compensation* (MHMC) is used to combine the first technique with a *boundary matching* (BM) temporal concealment technique to obtain a more robust performance.

The chapter also investigates the performance of different temporal error concealment techniques when incorporated within a multiple-reference video codec. In particular, the chapter finds a combination of techniques that best recovers the spatial-temporal components of a damaged multiple-reference motion vector. In addition, the chapter describes the design of a novel multihypothesis temporal concealment technique that can be used with multiple-reference systems.

The rest of the chapter is organized as follows. Section 10.2 describes the MFI temporal concealment technique, whereas Section 10.3 presents the combined BM-MFI technique. Section 10.4 presents some simulation results. Section 10.5 investigates the performance of temporal error concealment within multiple-reference video codecs. It also describes the multihypothesis multiple-reference temporal concealment technique. The chapter concludes with a discussion in Section 10.6.

Preliminary results of this chapter have appeared in Refs. 204, 205, 206, 207, and 208.

10.2   Temporal Error Concealment Using Motion Field Interpolation (MFI)

10.2.1   Motivation

As described earlier, conventional temporal concealment techniques estimate one concealment displacement for the whole damaged block and then use translational displacement compensation to conceal the block from a reference frame. As already discussed in Section 9.7.2, there are many cases where conventional temporal concealment techniques can fail and the concealment displacement can be incorrectly estimated. The main problem with such techniques is that incorrect estimation of the concealment displacement can lead to poor concealment of the entire or most of the block. This section describes a new temporal error concealment technique. This technique estimates one concealment displacement per pel of the damaged block and then conceals each pel individually. In this case, incorrect estimation of a concealment displacement will affect only the corresponding pel. On a block level, this may affect few pels rather than the entire block.

The described technique uses *motion field interpolation* (MFI) in its displacement estimation stage. In MFI, motion information needs to be available only at a number of nodal or control points within the motion field. The motion vector at any other point within the field can be approximated by interpolating the motion vectors of the surrounding control points. Thus, motion information recovery is inherent in MFI. As discussed in Chapter 5, MFI is used in warping-based motion compensation. Its main advantage over conventional translational compensation is that it provides a smoothly varying motion field that reduces blocking artefacts and compensates for more types of motion. These two features, i.e., *inherent motion information recovery* and *better motion compensation*, can improve both stages of the temporal concealment process, i.e., estimation and compensation, respectively. This makes MFI a very attractive choice for temporal error concealment.

10.2.2   Description of the Technique

Let $f_t(x, y)$ be the value of the current frame at pel location (x, y) and $f_{t-\Delta t}$ be a previously reconstructed and concealed frame. Further, let $\mathscr{D} = \{f_t(x, y): x \in [x_l, x_h], y \in [y_l, y_h]\}$ be a damaged block within the current frame and $\mathbf{v}_L$, $\mathbf{v}_R$, $\mathbf{v}_T$, and $\mathbf{v}_B$ be the motion vectors of the blocks to the left of, to the right of, above, and below the damaged block, respectively. The concealment displacement, $\hat{\mathbf{v}}(x, y) = (\hat{v}_x(x, y), \hat{v}_y(x, y))$, at any pel (x, y) within the damaged block $\mathscr{D}$ can be estimated by interpolating the neighboring motion vectors as follows:

$$\hat{\mathbf{v}}(x, y) = \frac{h_\gamma(x_n)\mathbf{v}_L + (1 - h_\gamma(x_n))\mathbf{v}_R + h_\gamma(y_n)\mathbf{v}_T + (1 - h_\gamma(y_n))\mathbf{v}_B}{2}, \quad (10.1)$$

$$y_n = \frac{y - y_l}{y_h - y_l}, \qquad x_n = \frac{x - x_l}{x_h - x_l}, \quad (10.2)$$

where (x_n, y_n) are the normalized spatial coordinates of pel (x, y) within the damaged block, ranging from $(0, 0)$ at the top-left corner to $(1, 1)$ at the bottom-right corner, and $h_\gamma(\cdot)$ is a suitable *interpolation kernel*.

Thus, the estimated displacement is a weighted sum of the neighboring motion vectors. The interpolation kernel, $h_\gamma(\cdot)$, is used to adjust the weights according to the spatial location of the pel within the damaged block. Intuitively, a pel on the left border should have a high contribution from the left vector, $\mathbf{v}_L$, and a low contribution from the right vector, $\mathbf{v}_R$, and so on. To achieve this, the following interpolation kernel [209] was used:

$$h_\gamma(a) = \frac{k(\gamma(2a - 1)) - k(\gamma)}{k(-\gamma) - k(\gamma)}, \quad 0 \leq a \leq 1 \text{ and } \gamma \geq 1, \quad (10.3)$$

where

$$k(b) = \frac{1}{1 + e^b}. \tag{10.4}$$

The parameter γ in Equation (10.3) is used to control the smoothness of the interpolation kernel. As γ varies from 1 to ∞, the interpolation kernel varies from an approximately linear shape to a brickwall shape, as illustrated in Figure 10.1.

Once the concealment displacement is estimated, then the damaged pel is concealed as follows:

$$\hat{f}_t(x, y) = f_{t-\Delta t}(x + \hat{v}_x(x, y), y + \hat{v}_y(x, y)). \tag{10.5}$$

In the case where the estimation process produces a subpel accurate displacement, the compensation process will require accessing a pel at a nonsampling location within the reference frame. Interpolation (e.g., bilinear) of the pels at surrounding sampling locations can be employed to provide an approximation to the required pel.

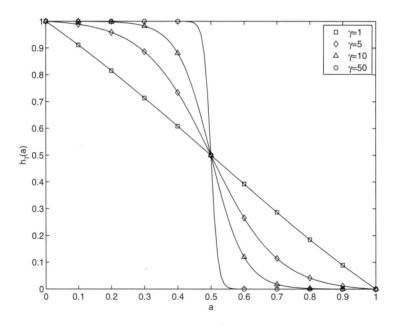

Figure 10.1: Interpolation kernel $h_\gamma(\cdot)$ with different values of the smoothness parameter γ

Table 10.1: Computational complexity of the displacement estimation stage of different temporal concealment techniques with a block of 16×16 pels

	Add/subtract	Multiply/divide	Magnitude
TR	—	—	—
AV	6	2	—
BM	496	—	256
MFI	**516**	**6**	—

10.2.3 Reduced-Complexity MFI

One of the main disadvantages of MFI is its high computational complexity. In the case of a linear interpolation kernel, Equation (10.1) reduces to

$$\hat{\mathbf{v}}(x, y) = \frac{(1 - x_n)\mathbf{v}_L + x_n\mathbf{v}_R + (1 - y_n)\mathbf{v}_T + y_n\mathbf{v}_B}{2}. \qquad (10.6)$$

A direct implementation of Equations (10.6) and (10.2) requires $10N^2$ additions/subtractions and $12N^2$ multiplications/divisions for an $N \times N$ block. This complexity can be reduced using a number of methods. One method is to calculate the weights off-line and store them in a lookup table. This reduces the complexity to $6N^2$ additions/subtractions and $8N^2$ multiplications/divisions. Another method is to use a *line-scanning* technique. That is, once $\hat{\mathbf{v}}(x, y)$ is calculated, the displacement of the next pel in the row and the next pel in the column can be calculated as follows:

$$\hat{\mathbf{v}}(x + 1, y) = \hat{\mathbf{v}}(x, y) + \frac{\mathbf{v}_R - \mathbf{v}_L}{2N} \quad \text{and} \quad \hat{\mathbf{v}}(x, y + 1) = \hat{\mathbf{v}}(x, y) + \frac{\mathbf{v}_B - \mathbf{v}_T}{2N}. \quad (10.7)$$

It is very simple to derive Equations (10.7) from Equation (10.6). Note that the second term in both of Equations (10.7) is a constant and needs to be calculated only once per block. This line-scanning technique further reduces the complexity to $(2N^2 + 4)$ additions/subtractions and six multiplications/divisions.

Table 10.1 compares the computational complexity of different temporal concealment techniques for a 16×16 block. The figures in the table refer to the complexity of the displacement estimation stage and do not include the complexity of the displacement compensation stage.[1] The figures for BM are based on four candidate motion vectors and SAD as the SMD measure. They do not include the complexity of sorting the SMDs and choosing the vector with the minimum SMD. Although the MFI technique has the highest number of multiplications/divisions, this increased complexity can be justified by

---

[1] For MFI, the displacement compensation stage is more complex, since it may involve interpolation.

improved concealment quality, as will be shown later. A point to note here is that MFI will be used only for damaged blocks. Thus, provided that the error rate is relatively low, this will not increase the complexity of the decoder considerably.

10.3 Temporal Error Concealment Using a Combined BM-MFI Technique

10.3.1 Motivation

In this section, *multihypothesis motion compensation* (MHMC) [106] is used to further improve the second stage, i.e., compensation, of the temporal concealment process. In MHMC, a block is compensated using a weighted average of several motion-compensated predictions (hypotheses). This is a general term that can be used to describe techniques like overlapped motion compensation, bidirectional motion compensation, and any other technique that compensates individual pels using more than one motion vector. When applied to temporal error concealment, this means that each pel of the damaged block will be concealed using more than one concealment displacement. In the described technique, *two* concealment displacements are used per pel: one is estimated using BM, as described in Section 9.7.2, and the other is estimated using MFI, as described in Section 10.2. The BM technique was chosen because it is one of the best conventional temporal error concealment techniques. A similar combination between BM and overlapped motion compensation has also been reported in Ref. 198.

In addition to improving the second stage of the temporal concealment process, the combination of BM and MFI can provide a more robust performance. This can be explained as follows. There are many cases where the BM technique will fail but the MFI technique will not, and vice versa. In such cases, a combination may be more robust because it may average out the concealment distortion.

10.3.2 Description of the Technique

Let $\hat{\mathbf{m}}(x, y) = (\hat{m}_x(x, y), \hat{m}_y(x, y))$ be the displacement estimated using MFI to conceal pel (x, y) of the damaged block $\mathscr{D}$, and let $\hat{\mathbf{b}} = (\hat{b}_x, \hat{b}_y)$ be the displacement estimated using BM to conceal the whole block. Then pel (x, y) is concealed as follows:

$$\hat{f}_t(x, y) = w_\delta(x_n, y_n) f_{t-\Delta t}(x + \hat{m}_x(x, y), y + \hat{m}_y(x, y))$$
$$+ (1 - w_\delta(x_n, y_n)) f_{t-\Delta t}(x + \hat{b}_x, y + \hat{b}_y). \tag{10.8}$$

Thus, the concealed pel is a weighted sum of two predictions. The function $w_\delta(\cdot, \cdot)$ is used to adjust the weights given to MFI and BM according to the spatial location of the pel within the damaged block. Knowledge of the way both BM and MFI work can provide some insights into designing a suitable $w_\delta(\cdot, \cdot)$. For example, the SMD measure of the BM technique involves the border pels of the damaged block. It is expected, therefore, that BM will perform well at those pels. Therefore, BM must be given high weights at the borders of the block and low weights at the center. To achieve this, the following function was used:

$$w_\delta(x_n, y_n) = \frac{g_\delta(x_n)g_\delta(y_n) + 1}{2}, \qquad (10.9)$$

where

$$g_\delta(a) = \begin{cases} 1 - \frac{k(\delta(4a-1)) - k(\delta)}{k(-\delta) - k(\delta)}, & 0 \leq a \leq \frac{1}{2} \\ g_\delta(1-a), & \frac{1}{2} < a \leq 1 \end{cases} \qquad (10.10)$$

and $k(\cdot)$ is defined by Equation (10.4). The parameter δ is used to control the smoothness of $w_\delta(\cdot, \cdot)$, as illustrated in Figure 10.2.

Before proceeding to present simulation results, it is valuable at this point to highlight the main differences between the two novel algorithms, MFI and BM-MFI, and conventional temporal error concealment techniques. These are summarized in Table 10.2.

10.4 Simulation Results

10.4.1 Results Within an Isolated Error Environment

It is very important to evaluate the performance of the techniques in isolation from any external effects, like temporal and spatial error propagation and the choice of the error detection algorithm. This is particularly important for a fair comparison, since such error mechanisms and algorithm choices may randomly affect one technique more than another. Thus, in this set of simulations, the following assumptions were made

1. There is no temporal error propagation. This was achieved by using original reference frames for the concealment process.

2. There is no spatial error propagation. This is equivalent to using fixed-length codes and no predictive coding.

3. The concealment process is supported by an ideal error detection algorithm that can identify all damaged blocks.

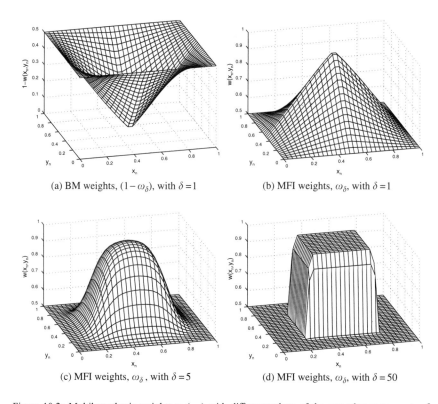

(a) BM weights, $(1 - \omega_\delta)$, with $\delta = 1$ (b) MFI weights, ω_δ, with $\delta = 1$

(c) MFI weights, ω_δ, with $\delta = 5$ (d) MFI weights, ω_δ, with $\delta = 50$

Figure 10.2: Multihypothesis weights $w_\delta(\cdot, \cdot)$ with different values of the smoothness parameter δ

Table 10.2: Comparison between conventional temporal concealment and the MFI and BM-MFI techniques

	Conventional temporal concealment	MFI	BM-MFI
Displacement estimation	One displacement per block using AV, TR, BM, etc.	One displacement per pel using MFI (weighted sum of four neighboring vectors)	Two displacements per pel: one produced by MFI and another produced by BM
Displacement compensation	Translational, same displacement for the whole block.	Translational, but on a pel-by-pel basis	Multihypothesis motion compensation (each pel is a weighted sum of two concealments)

Hereafter, the term *isolated error environment* will be used to refer to this set of test conditions.

All results in this subsection were generated using a full-search block-matching algorithm with blocks of 16×16 pels, a maximum allowed displacement of ± 15 pels, SAD as the distortion measure, restricted motion vectors, and full-pel accuracy. Block losses were introduced randomly. Five temporal error concealment techniques were simulated: temporal replacement (TR), average vector (AV), boundary matching with side-match distortion (BM), motion field interpolation (MFI), and the combination of BM and MFI (BM-MFI). In each technique, the motion vectors of the four neighboring blocks—left, right, above and below—were used in the concealment displacement estimation stage. Whenever a neighboring motion vector was not available, e.g., damaged or does not exist as in border blocks, it was set to $(0,0)$. For the BM technique, SAD was used in the side-match distortion calculations. Again, to mask any external effects, all quoted PSNRs in this set of simulations were calculated for concealed blocks only and averaged over the whole sequence. All quoted results refer to the luma components of sequences.

10.4.1.1 Choice of Parameters

Before evaluating the performance of MFI and BM-MFI, suitable values for the smoothness parameters γ and δ need to be chosen. Figure 10.3 shows the effect of changing the smoothness parameter γ on the performance of MFI when applied to FOREMAN at 25 frames/s with different block loss rates. In general, the performance is not particularly sensitive to the choice of γ (a change of about 0.3 dB). As γ increases, the performance of MFI deteriorates slightly. The best performance is achieved with $\gamma = 1$. This is approximately a linear kernel. Thus, a linear interpolation kernel will be used in all subsequent simulations. Note that a linear kernel also facilitates the use of a line-scanning technique to reduce complexity, as was shown in Section 10.2.3.

Figure 10.4 shows the effect of changing the smoothness parameter δ on the performance of BM-MFI when applied to FOREMAN at 25 frames/s with different block loss rates. Again, the performance is not very sensitive to changes in δ. As δ increases, the performance of BM-MFI slightly deteriorates. The best performance is achieved with $\delta = 1$. The corresponding multihypothesis weights are those shown in Figures 10.2(a) and 10.2(b). In what follows, this value of δ will be used.

10.4.1.2 Performance Evaluation

Figures 10.5, 10.6, and 10.7 compare the performance of the five techniques when applied to AKIYO, FOREMAN, and TABLE TENNIS, respectively. All results were generated with a frame skip of 1.

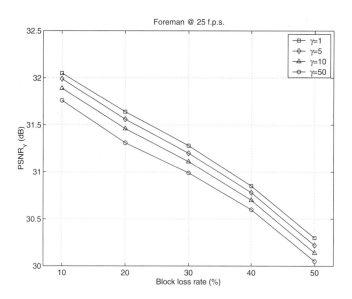

Figure 10.3: Performance of MFI when applied to QSIF FOREMAN at 25 frames/s with different interpolation kernels. PSNRs are for damaged blocks only

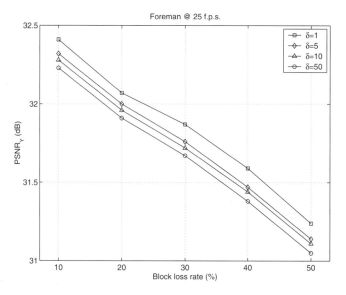

Figure 10.4: Performance of BM-MFI when applied to QSIF FOREMAN at 25 frames/s with different multihypothesis weights. PSNRs are for damaged blocks only

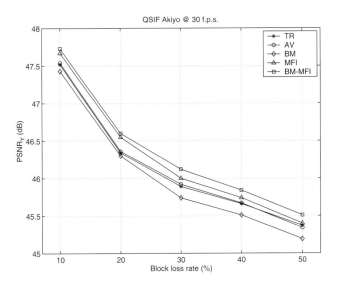

Figure 10.5: Comparison between different temporal concealment techniques when applied to QSIF Akiyo at 30 frames/s. PSNRs are for damaged blocks only

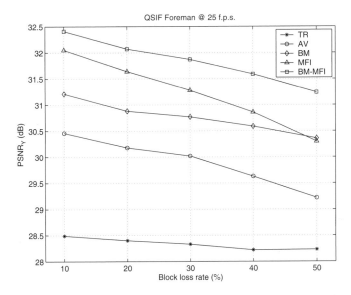

Figure 10.6: Comparison between different temporal concealment techniques when applied to QSIF Foreman at 25 frames/s. PSNRs are for damaged blocks only

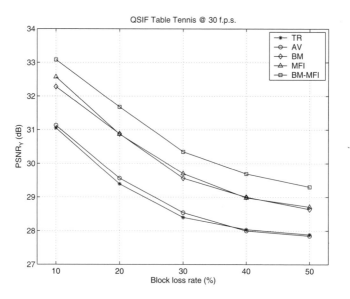

Figure 10.7: Comparison between different temporal concealment techniques when applied to QSIF TABLE TENNIS at 30 frames/s. PSNRs are for damaged blocks only

In general, the best performance was achieved by BM-MFI, followed by MFI, then BM, AV, and TR. As expected, TR performs well for the low-movement AKIYO sequence. The poor performance of BM for AKIYO may be due to an ambiguity problem where neighboring motion vectors give similar SMD measures. A very interesting point to note is that the performance of MFI starts to deteriorate for FOREMAN at high block loss rates. This may be due to the high dependency of MFI on the availability of the neighboring motion vectors. This can be improved using interleaving techniques, as was described in Chapter 9. In all cases, however, the BM-MFI technique maintained its superior performance. This is a clear indication of the robustness of the technique. Over the three sequences and the considered block loss rate range, MFI provides on average 0.3 dB, 0.9 dB, and 1.4 dB improvements over BM, AV, and TR, respectively, whereas BM-MFI provides a further 0.5 dB improvement over MFI. This corresponds to improvements of about 0.8 dB, 1.4 dB, and 1.9 dB over BM, AV, and TR, respectively.

Figure 10.8 shows the subjective quality of the 58th frame of TABLE TENNIS with a block loss rate of 30% when concealed using BM and BM-MFI. The superior performance of the BM-MFI technique is immediately evident from the good concealment of the left hand of the player. Note, however, that some parts of the shirt are less sharp with the BM-MFI technique. This may be due to the low-pass filtering effect of the averaging (weighting) process.

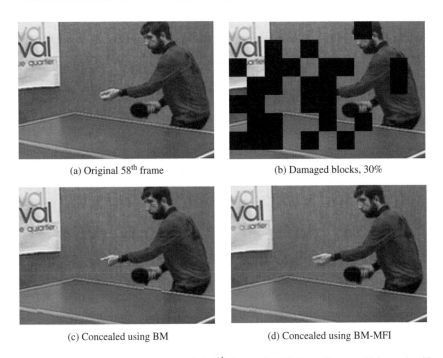

(a) Original 58th frame   (b) Damaged blocks, 30%

(c) Concealed using BM   (d) Concealed using BM-MFI

Figure 10.8: Subjective quality of concealed 58th frame of QSIF Table Tennis at 30 frames/s with a block loss rate of 30%

10.4.2 Results Within an H.263 Decoder

This set of simulations tests the performance of the techniques when incorporated within an H.263 decoder. In this case, the assumptions made in the previous set of simulations will be relaxed. In other words, previously reconstructed, possibly damaged and concealed, frames will be used for both prediction and concealment. This will result in temporal error propagation. In addition, spatial error propagation will also occur, since H.263 uses VLC and predictive coding.

The Telenor implementation [144] of H.263 was used in this simulation. The decoder was modified to perform error detection by detecting syntax and semantic violations, as was described in Section 9.5. When an error is detected, the decoding process is stopped, the decoder searches for the next synchronization codeword, and decoding is resumed. All macroblocks between the point where the error was detected and the synchronization point are marked as damaged macroblocks. In this simulation, the H.263 encoder option to insert synchronization codewords at the start of each GOB was switched on. All

Table 10.3: Comparison between different temporal concealment techniques when applied to three test sequences corrupted with a random bit error rate of 10^{-3}. PSNRs are for whole frames

		Error free	TR	AV	BM	MFI	BM-MFI
AKIYO	$PSNR_{Y'}$	35.15	30.01	29.93	28.19	30.21	30.35
12 kbits/s	$PSNR_{C'_R}$	37.18	34.01	33.67	30.61	34.12	34.23
	$PSNR_{C'_B}$	39.16	36.14	36.03	35.00	36.20	36.32
	PSNR	**35.92**	**31.12**	**31.02**	**29.18**	**31.31**	**31.45**
FOREMAN	$PSNR_{Y'}$	27.93	19.11	19.30	19.59	19.56	20.05
24 kbits/s	$PSNR_{C'_R}$	35.02	30.49	29.01	28.92	30.64	30.67
	$PSNR_{C'_B}$	34.54	29.93	29.37	29.23	30.34	30.40
	PSNR	**29.26**	**20.71**	**20.84**	**21.11**	**21.15**	**21.62**
TABLE TENNIS	$PSNR_{Y'}$	33.21	18.36	18.25	18.58	18.68	18.95
48 kbits/s	$PSNR_{C'_R}$	38.21	23.86	22.50	22.40	23.91	23.93
	$PSNR_{C'_B}$	36.79	21.79	21.32	21.42	22.22	22.34
	PSNR	**34.22**	**19.39**	**19.16**	**19.43**	**19.70**	**19.94**

other optional modes were switched off. No INTRA refresh was employed. Thus, only the first frame was INTRA coded.

The H.263 encoder was used to encode the three sequences AKIYO, FOREMAN, and TABLE TENNIS[2] at bit rates of 12 kbits/s, 24 kbits/s, and 48 kbits/s, respectively. Note that the bit rates were chosen according to the amount of spatial detail and movement within each sequence. Note also that all bit rates were chosen within the very-low-bit-rate range, i.e., less than 64 kbits/s.

The compressed bitstreams were corrupted with random bit errors generated according to the MPEG-4 error robustness test specification [210]. The specifications provide an initial period of 1.5 s during which no errors are injected. This allows for the encoder to transmit an initial INTRA frame and for the codec operation to stabilize into a steady state before errors are introduced.

Table 10.3 summarizes the performance of the five techniques when applied to the three test sequences with a frame skip of 1 and a bit error rate (BER) of 10^{-3}. The quoted PSNRs are for whole frames and averaged over the sequence. The quantities $PSNR_{Y'}$, $PSNR_{C'_R}$, and $PSNR_{C'_B}$ represent the PSNRs of the separate luma and two chroma components, respectively, whereas PSNR

---

[2]The luma components of both AKIYO and TABLE TENNIS were zero-padded vertically to 128 lines because Telenor's H.263 can work only with an integer multiple of 16. The corresponding chroma components were also appropriately padded.

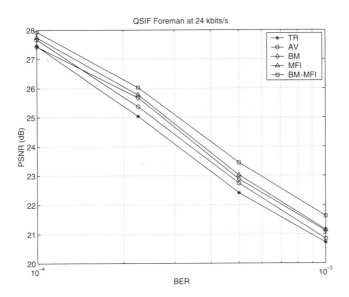

Figure 10.9: Comparison between different temporal concealment techniques when applied to QSIF FOREMAN. The sequence was H.263 encoded at 24 kbits/s and then corrupted with a range of bit error rates

represents the PSNR of the three components together with a 4:2:0 subsampling. Again, the best performance in each case was achieved by BM-MFI, followed by MFI. For example, for the TABLE TENNIS sequence, MFI provides improvements of 0.27 dB, 0.54 dB, and 0.31 dB over BM, AV, and TR, respectively, whereas BM-MFI provides a further 0.24 dB improvement over MFI. This corresponds to improvements of about 0.51 dB, 0.78 dB, and 0.55 dB over BM, AV, and TR, respectively.

Figure 10.9 shows the performance of the five techniques when used to conceal the 24 kbits/s QSIF FOREMAN sequence corrupted with BERs in the range 10^{-4} to 10^{-3}. At low BERs the differences between the techniques are small. However, as the BER increases, the techniques split into three performance levels. The lowest level includes TR and AV, the next level includes BM and MFI, and the highest level includes BM-MFI.

Figure 10.10 shows a frame of the 24 kbits/s QSIF FOREMAN corrupted with a BER of 10^{-3} and then decoded and concealed using BM and BM-MFI. The superior performance of the BM-MFI technique is immediately evident, especially at the eyes and the edges of the face.

It is worth noting here that at a BER of 10^{-3}, the PSNRs of the concealed sequences drop by about 5–9 dB compared to the error-free values, and the subjective quality may not be acceptable. A close inspection of the

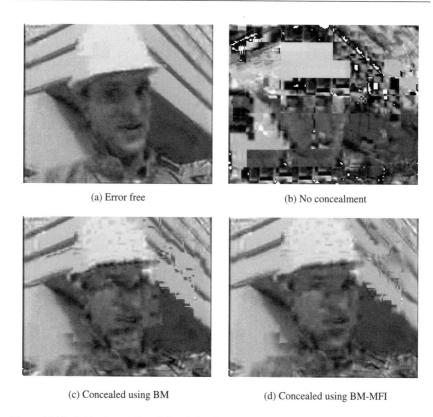

(a) Error free   (b) No concealment

(c) Concealed using BM   (d) Concealed using BM-MFI

Figure 10.10: Subjective quality of decoded and concealed frame of QSIF FOREMAN. The sequence was H.263 encoded at 24 kbits/s and corrupted with a 10^{-3} bit error rate

decoded and concealed sequences revealed that this poor performance is due mainly to the effects of spatial and temporal error propagation and also to the imperfections of the error detection approach. In addition, it was observed that temporal techniques do not perform well for intracoded blocks, scene changes, and uncovered backgrounds. Thus, despite their advantages, temporal error concealment techniques must be combined with spatial error concealment and, more importantly, must be supported by some error containment techniques, such as INTRA refresh.

10.5   Temporal Error Concealment for Multiple-Reference Motion-Compensated Prediction

As already discussed, temporal error concealment is an important tool to combat the effects of errors on transmitted video. A number of temporal error concealment techniques have been proposed in the literature, and their performances have been extensively studied within typical single-reference video codecs operating over various error-prone channels. There is, however, a *need* to characterize the performance of such techniques within multiple-reference video codecs. This is the main aim of this section.

Temporal error concealment within a multiple-reference video codec can be split into two problems: *spatial-components* (d_x, d_y) *recovery* and *temporal-component* d_t *recovery*. Thus, a multiple-reference temporal error concealment method can be represented by a combination of the form S-T, where S is the technique used to recover the spatial components and T is the technique used to recover the temporal component. In this section, S and T can be chosen from the following list of techniques

ZR The recovered motion component (either spatial or temporal) is set to *zero*. In Chapter 9 this was referred to as temporal replacement (TR).

AV The recovered motion component is set to the *average* of the *corresponding* components of a set of neighboring motion vectors. In this section, four neighboring vectors are used: top, bottom, left, and right.

BM This is a boundary-matching method (refer to Section 9.7.2 for a detailed description). A set of candidate vectors is first chosen. Each candidate is then used to conceal the damaged block. The quality of this concealment is assessed using the side-match distortion (SMD) measure, which is defined as the sum of absolute (or squared) differences across the four boundaries of the block. The candidate with the minimum SMD is chosen. In this section, the set of candidates includes the four neighboring vectors—top, bottom, left, and right—and the SMD is defined as the SAD across the boundaries.

MFI This is the method described in this chapter. It uses motion field interpolation to recover one vector per pel of the damaged block. In this section a linear interpolation kernel is employed.

Since there are four techniques in the list, there are 16 possible combinations of the form S-T. Each combination leads to a different long-term temporal concealment method. For example, assume that $\mathbf{l} = (l_x, l_y, l_t)$, $\mathbf{r} = (r_x, r_y, r_t)$, $\mathbf{t} = (t_x, t_y, t_t)$, and $\mathbf{b} = (b_x, b_y, b_t)$ are, respectively, the motion vectors of the blocks to the left of, to the right of, above, and below the damaged block. A combination of the form AV-BM means that the spatial components (d_x, d_y) are first recovered using the AV method:

$$\hat{d}_x = \frac{l_x + r_x + t_x + b_x}{4} \quad \text{and} \quad \hat{d}_y = \frac{l_y + r_y + t_y + b_y}{4}. \tag{10.11}$$

Then a set $\mathscr{C} = \{\mathbf{d}_1, \ldots, \mathbf{d}_4\}$ of four candidates is formed from the recovered spatial components $(\hat{d}_x, \hat{d}_y)$ and the four temporal components of the neighboring blocks. In other words: $\mathbf{d}_1 = (\hat{d}_x, \hat{d}_y, l_t)$, $\mathbf{d}_2 = (\hat{d}_x, \hat{d}_y, r_t)$, $\mathbf{d}_3 = (\hat{d}_x, \hat{d}_y, t_t)$, and $\mathbf{d}_4 = (\hat{d}_x, \hat{d}_y, b_t)$. The BM technique is then used to recover the temporal component by choosing from this set of candidates. Thus

$$\hat{\mathbf{d}} = \arg \min_{\mathbf{d}_i \in \mathscr{C}} \mathrm{SMD}(\mathbf{d}_i). \qquad (10.12)$$

A multiple-reference rate-constrained H.263-like codec was used to generate the results of this section. This codec uses full-pel full-search block matching with macroblocks of 16×16 pels, a maximum allowed spatial displacement of ± 15 pels, SAD as the distortion measure, restricted motion vectors, and reconstructed reference frames. Motion vectors are coded using the median predictor and the VLC table of the H.263 standard. The frame signal (in case of INTRA) and the DFD signal (in case of INTER) are transform encoded according to the H.263 standard. The codec uses rate-constrained motion estimation and mode decision as defined in the high-complexity mode of TMN10. The codec employs a sliding-window control to maintain a long-term memory of size $M = 10$ frames. Only the first frame is INTRA coded, and no INTRA refresh is employed. A fixed quantization parameter of $QP = 10$ is used. Errors were introduced randomly on a macroblock level. Thus, an error rate of 20% means that 20% of the macroblocks are damaged per frame. It is assumed that the decoder uses an ideal error detection mechanism. All quoted results refer to the luma components of sequences.

10.5.1 Temporal-Component Recovery

This set of experiments investigate the best technique for recovering the temporal component d_t of a damaged long-term motion vector. In this case, the spatial recovery technique S, in the combination S-T, was kept constant at ZR, whereas the temporal recovery technique T was varied over ZR, AV, BM, and MFI. In other words, four S-T combinations were considered: ZR-ZR, ZR-AV, ZR-BM, and ZR-MFI.

Figures 10.11, 10.12, and 10.13 show the results for the QSIF sequences AKIYO, FOREMAN, and TABLE TENNIS, respectively. Part (a) of each figure shows the performance with a frame skip of 3 over a range of macroblock error rates, whereas part (b) shows the performance with a macroblock error rate of 20% over a range of frame skips.

In general, the best temporal-component recovery is achieved by ZR and BM (i.e., ZR-ZR and ZR-BM). The good performance of ZR is due to the zero-biased distribution of the temporal components (Property 6.3.1.2). In other words, the temporal component $d_t = 0$ has the highest frequency of occurrence

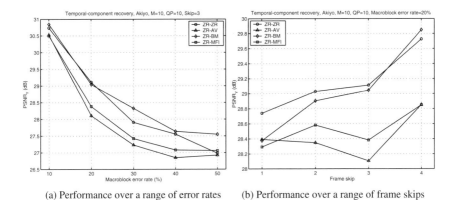

(a) Performance over a range of error rates   (b) Performance over a range of frame skips

Figure 10.11: Temporal-component recovery for QSIF Akiyo with $M = 10$ and $QP = 10$

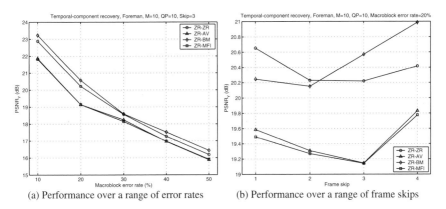

(a) Performance over a range of error rates   (b) Performance over a range of frame skips

Figure 10.12: Temporal-component recovery for QSIF Foreman with $M = 10$ and $QP = 10$

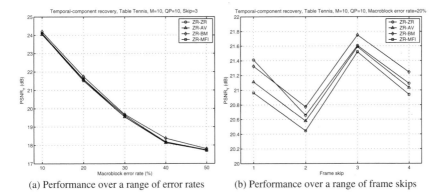

(a) Performance over a range of error rates   (b) Performance over a range of frame skips

Figure 10.13: Temporal-component recovery for QSIF Table Tennis with $M = 10$ and $QP = 10$

within the long-term memory block-motion field. Note that at low frame skips, this simple ZR method is sufficient, whereas at high frame skips the more complex method of BM has to be employed. This may be due to the fact that at high frame skips, the zero-biased distribution becomes more spread (see Property 6.3.1.2 and Figure 6.3). In other words, $d_t = 0$ becomes less probable, and longer temporal components start to appear more frequently in the motion field. Such components need to be recovered using BM. Both AV and MFI provide poor temporal-component recovery compared to BM and ZR.

10.5.2   Spatial-Components Recovery

This set of experiments investigates the best technique for recovering the spatial components (d_x, d_y) of a damaged long-term motion vector. In this case, the temporal recovery technique T, in the combination S-T, was kept constant at ZR, whereas the spatial recovery technique S was varied over ZR, AV, BM, and MFI. In other words, four S-T combinations were considered: ZR-ZR, AV-ZR, BM-ZR, and MFI-ZR.

Figures 10.14, 10.15, and 10.16 show the results for the QSIF sequences AKIYO, FOREMAN, and TABLE TENNIS, respectively. Part (a) of each figure shows the performance with a frame skip of 3 over a range of macroblock error rates, whereas part (b) shows the performance with a macroblock error rate of 20% over a range of frame skips.

In general, the best spatial-components recovery is achieved by MFI followed by BM. This is similar to the single-reference results reported in Section 10.4. Thus, moving from a single-reference system to a multiple-reference system does not significantly influence the spatial-components

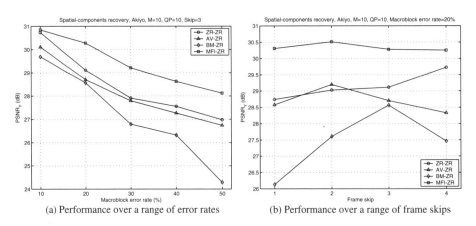

(a) Performance over a range of error rates   (b) Performance over a range of frame skips

Figure 10.14: Spatial-components recovery for QSIF AKIYO with $M = 10$ and QP $= 10$

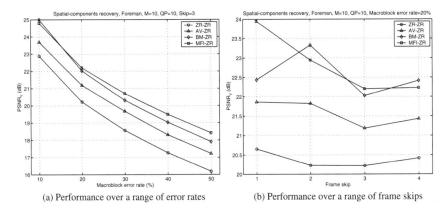

(a) Performance over a range of error rates   (b) Performance over a range of frame skips

Figure 10.15: Spatial-components recovery for QSIF FOREMAN with $M = 10$ and $QP = 10$

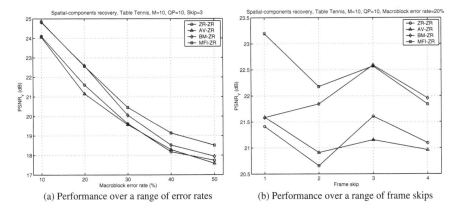

(a) Performance over a range of error rates   (b) Performance over a range of frame skips

Figure 10.16: Spatial-components recovery for QSIF TABLE TENNIS with $M = 10$ and $QP = 10$

recovery process. A very interesting point to note is that the performance of MFI starts to deteriorate at high frame skips. This may be due to the fact that at high frame skips, the spatial components within the motion field become less correlated (see Property 6.3.1.3 and Figures 6.4(a) and 6.4(b)). Since MFI assumes a high correlation between the spatial components, its performance will deteriorate with decreased correlation.

10.5.3 Spatial-Temporal-Components Recovery

Comparing the results of Section 10.5.1 to those of Section 10.5.2 it can be concluded that spatial-components recovery is, in general, more important than

Table 10.4: Spatial-temporal recovery for QSIF AKIYO with $M = 10$, QP $= 10$, skip $= 3$, and a macroblock error rate of 30%

		Spatial-components recovery			
		ZR	AV	BM	MFI
Temporal-	TR	27.91	27.80	26.79	29.22
component	AV	27.23	27.49	27.07	28.58
recovery	BM	28.33	28.10	27.38	**29.48**
	MFI	27.42	27.58	26.38	28.69

Table 10.5: Spatial-temporal recovery for QSIF FOREMAN with $M = 10$, QP $= 10$, skip $= 3$, and a macroblock error rate of 30%

		Spatial-components recovery			
		ZR	AV	BM	MFI
Temporal-	TR	18.57	19.68	20.32	20.71
component	AV	18.25	19.58	19.72	20.56
recovery	BM	18.59	20.13	20.80	**21.18**
	MFI	18.14	19.51	19.65	20.48

temporal-component recovery. For example, in Figure 10.13(b), at a frame skip of 3, moving from the best technique, ZR-BM, to the worst technique, ZR-MFI, drops the quality by about 0.3 dB, whereas in Figure 10.16(b) moving from the best technique, MFI-ZR, to the worst technique, AV-ZR, drops the quality by about 1 dB. It can be concluded also that spatial-components recovery is, in general, more complex than temporal-component recovery. With temporal-component recovery, a simple technique like ZR can be sufficient, whereas with spatial-components recovery more complex techniques like MFI and BM are essential. Furthermore, the results of Sections 10.5.1 and 10.5.2 indicate that the combination MFI-BM (i.e., spatial recovery using MFI and temporal recovery using BM) may provide the best spatial-temporal recovery. This is confirmed in Tables 10.4, 10.5, and 10.6, which show the performance of all 16 possible combinations with a frame skip of 3 and a macroblock error rate of 30%.

10.5.4 Multihypothesis Temporal Error Concealment

It was demonstrated in Section 10.4 that a more robust performance can be achieved if the concealed block is a weighted average of a number of

Table 10.6: Spatial-temporal recovery for QSIF TABLE TENNIS with $M = 10$, $QP = 10$, skip $= 3$, and a macroblock error rate of 30%

		Spatial-components recovery			
		ZR	AV	BM	MFI
Temporal-	TR	19.62	19.57	20.06	20.46
component	AV	19.54	19.57	20.02	20.40
recovery	BM	19.68	19.87	20.09	**20.58**
	MFI	19.55	19.74	20.00	20.40

candidate concealments, where each candidate concealment is provided using a different recovered motion vector. This is very similar to multihypothesis motion compensation [106]. Thus, it is termed *multihypothesis temporal error concealment*.

In this subsection a multihypothesis temporal concealment technique to be used with long-term memory motion-compensated prediction is presented. In this case, the candidate concealments are taken from different reference frames. The details of this technique are as follows. The spatial components are first recovered using MFI (as suggested in Section 10.5.2). However, instead of recovering a single temporal component, all four neighboring temporal components are utilized. Combined with the recovered spatial components, each neighboring temporal component provides a candidate concealment from the corresponding reference frame. The four candidate concealments are then averaged and used to conceal the damaged block in the current frame. In other words, a damaged pel (x, y) in the current frame f_c is concealed as follows:

$$\hat{f}_c(x, y) = \frac{1}{4} \sum_{i=1}^{4} f_r(x + \hat{d}_x(x, y), y + \hat{d}_y(x, y), d_{t_i}), \qquad (10.13)$$

where $f_r(\cdot, \cdot, d_t)$ refers to reference frame d_t in the multiframe memory, $(\hat{d}_x(x, y), \hat{d}_y(x, y))$ are the spatial components recovered at pel (x, y) using MFI, and d_{t_i}, $i = 1, \ldots, 4$ are the temporal components of the four neighboring vectors. In what follows, this approach is designated as MFI-MH.

Figures 10.17, 10.18, and 10.19 compare the performance of the MFI-MH technique to that of MFI-BM (which is the best combination, as suggested in Section 10.5.3) and also to that of ZR-ZR (which is the simplest and most commonly used combination). The figures confirm the superior performance of the suggested combination, MFI-BM, compared to the most commonly used combination, ZR-ZR. In addition, the figures show that further improvements can be achieved using the multihypothesis MFI-MH technique.

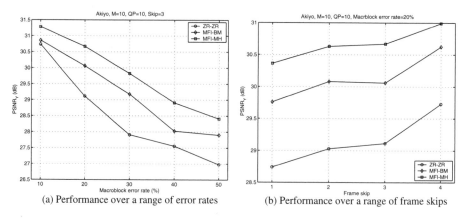

(a) Performance over a range of error rates   (b) Performance over a range of frame skips

Figure 10.17: Multihypothesis temporal concealment for QSIF A$_{\text{KIYO}}$ with $M = 10$ and $QP = 10$

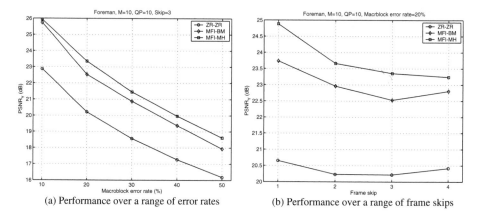

(a) Performance over a range of error rates   (b) Performance over a range of frame skips

Figure 10.18: Multihypothesis temporal concealment for QSIF F$_{\text{OREMAN}}$ with $M = 10$ and $QP = 10$

This is also confirmed using Figure 10.20, which shows the subjective quality of the 102nd frame of QSIF F$_{\text{OREMAN}}$ encoded at 8.33 frames/s with $M = 10$, $QP = 10$, and corrupted with a random macroblock error rate of 20%. Figure 10.20(a) shows the error-free reconstructed frame, whereas Figure 10.20(b) shows the locations of the damaged macroblocks in addition to errors propagated from previous frames. Figures 10.20(c), 10.20(d), and 10.20(e) show the same frame when concealed using ZR-ZR, MFI-BM, and MFI-MH, respectively. The figures clearly show that the suggested MFI-BM combination and the multihypothesis MFI-MH technique both outperform the commonly used ZR-ZR technique. In addition, the figures clearly show the superior subjective quality of the MFI-MH technique (Figure 10.20(e)), even

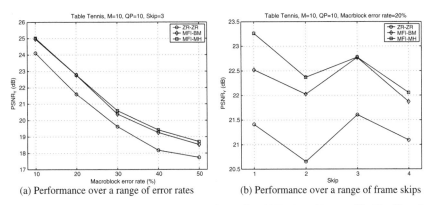

(a) Performance over a range of error rates   (b) Performance over a range of frame skips

Figure 10.19: Multihypothesis temporal concealment for QSIF TABLE TENNIS with $M = 10$ and $QP = 10$

over that of MFI-BM (Figure 10.20(d)). In particular note the left eye of Foreman (to the right of the viewer) and the diagonal lines in the walls.

10.6   Discussion

Because of their simplicity, no added redundancy, and minimum delay, error concealment techniques were identified in this chapter as the most suitable techniques for mobile video applications. Thus, it was decided to concentrate on error concealment and in particular on temporal techniques.

Conventional temporal concealment techniques estimate one concealment displacement for the whole damaged block and then use translational displacement compensation to conceal the block from a reference frame. It was realized, therefore, that wrong estimation of the concealment displacement can lead to poor concealment of the entire or most of the block. To overcome this drawback, a novel temporal concealment technique was designed. In this technique, motion field interpolation (MFI) is used to estimate one concealment displacement per pel of the damaged block. Each pel is then concealed individually. In this case, incorrect estimation of a concealment displacement will affect only the corresponding pel rather than the entire block. The inherent motion information recovery and the good motion compensation performance of the MFI technique improve both stages of temporal concealment, i.e., estimation and compensation.

To achieve a more robust performance, a second novel temporal concealment technique was also designed. In this technique, multihypothesis motion compensation (MHMC) is used to combine the MFI technique with a

(a) Error free (32.26 dB)   (b) Locations of errors (with propagation)

(c) Concealed using ZR-ZR (20.62 dB)   (d) Concealed using MFI-BM (22.91 dB)

(e) Concealed using MFI-MH (25.05 dB)

Figure 10.20: Subjective quality of 102nd frame of QSIF FOREMAN encoded at 8.33 frames/s with $M = 10$, $QP = 10$, and corrupted with a macroblock error rate of 20%

boundary matching (BM) temporal technique. In effect, this improves the second stage of temporal concealment, i.e., compensation.

Simulation results, within both an isolated error environment and an H.263 codec, showed the superior objective and subjective performances of the designed techniques. The MFI technique achieved reasonable improvements over conventional temporal concealment techniques, but it was found that its performance can slightly deteriorate at very high error rates. The combined BM-MFI technique showed a more superior and robust performance at all error rates.

It was also observed that factors like spatial and temporal error propagation, imperfections of the error detection algorithm, scene changes, and uncovered background can severely degrade the performance of temporal concealment techniques. Thus, despite their advantages, such techniques must be combined with spatial techniques and must also be supported by powerful error detection and error containment techniques.

The chapter also investigated the performance of temporal error concealment techniques when incorporated within an LTM-MCP codec. It was found that the best techniques to recover the temporal component are zero replacement (ZR) and boundary matching (BM). The former is sufficient at low frame skips, whereas the latter is preferred at high frame skips. It was also found that the best technique to recover the spatial components is the MFI technique. All these findings were explained in view of the properties of the long-term memory block-motion field. In general, it was concluded that spatial-components recovery is more complex and more important than temporal-component recovery. In addition, a combination of the form MFI-BM (i.e., spatial recovery using MFI and temporal recovery using BM) will provide the best spatial-temporal recovery. In order to achieve a more robust performance, the chapter described the design of a multihypothesis multiple-reference temporal concealment technique. In this technique, a damaged block is concealed using the average of four candidate concealments, probably from different reference frames. Simulation results showed the superior performance of this technique.

Appendix

Fast Block-Matching Algorithms

A.1 Notation and Assumptions

- BDM: a block distortion measure, like the SSD or SAD.

- d_m: maximum allowed motion displacement.

- N: total number of steps in the search. It is an integer number greater than 0.

- s: current search step size.

- (c_x, c_y): current search center.

- (m_x, m_y): current location of minimum distortion.

- (d_x, d_y): final motion vector.

- $\lfloor \cdot \rfloor$: floor operator. It rounds its argument to the nearest integer toward $-\infty$.

- $\lceil \cdot \rceil$: ceil operator. It rounds its argument to the nearest integer toward $+\infty$.

- min: minimize operator. It returns the minimum of a given function.

- max: maximize operator. It returns the maximum of a given function.

- arg: argument operator. It returns the argument of a given function.

- All algorithms presented in this appendix assume full-pel accuracy. Sub-pel accuracy can easily be achieved using very minor modifications.

- If the search procedure attempts to search a location outside the search window, the corresponding BDM is set to a maximum value.

- It is assumed that the search procedure keeps a record of all locations searched so far and their BDM values. This avoids reevaluating the same BDMs in subsequent steps.

A.2   The Two-Dimensional Logarithmic (TDL) Search

The two-dimensional logarithmic (TDL) search was proposed by Jain and Jain in 1981 [54]. It uses a uniform search pattern of five locations (the center and endpoints of a + shape). At each step, the search pattern is centered at the minimum location from the previous step. The step size is halved if the center of the search is the same as that of the previous step. The search is stopped when the step size is 1. In this case, nine locations, rather than five, are searched (the center and endpoints of a * shape) to find the final motion vector. The TDL algorithm is described in the following procedure.

1. Initialize the search step size to

$$s = \max(2, 2^{\lfloor \log_2 d_m \rfloor - 1}).$$

2. Initialize the center of search to the origin of the search window:

$$(c_x, c_y) = (0, 0).$$

3. Evaluate the BDM at the center of the search and its four vertical and horizontal neighbors at a step size of s. Out of this set of five locations, find the one that achieves the minimum BDM:

$$(m_x, m_y) = \arg \min_{(i,j) \in \mathcal{P}_1} \text{BDM}(i, j),$$

where

$$\mathcal{P}_1 = \{(c_x, c_y), (c_x + s, c_y), (c_x - s, c_y), (c_x, c_y + s), (c_x, c_y - s)\}.$$

4. IF the minimum is at the center of the search pattern, i.e., if $(m_x, m_y) = (c_x, c_y)$, THEN

   (a) Halve the search step size:

   $$s = \frac{s}{2}.$$

   (b) IF the search step size is 1, i.e., if $s = 1$, THEN

       i. Evaluate the BDM at the center of the search and its eight immediate neighbors. Out of this set of nine locations, set the motion vector to the one that achieves the minimum BDM:

       $$(d_x, d_y) = \arg \min_{(i,j) \in \mathcal{P}_2} \text{BDM}(i, j),$$

where

$$\mathscr{P}_2 = \{(c_x, c_y), (c_x + 1, c_y), (c_x - 1, c_y), (c_x, c_y + 1), (c_x, c_y - 1),$$
$$(c_x - 1, c_y - 1), (c_x - 1, c_y + 1), (c_x + 1, c_y - 1), (c_x + 1, c_y + 1)\}.$$

   ii. STOP

(c) ELSE (when the step size is not 1, i.e., $s \neq 1$) GOTO step 3.

5. ELSE (when the minimum is not in the center, i.e., $(m_x, m_y) \neq (c_x, c_y)$)

(a) Set the center of the search to the new minimum location:

$$(c_x, c_y) = (m_x, m_y).$$

(b) GOTO step 3.

A.3   The N-Steps Search (NSS)

This is the general form of the three-steps search (TSS) reported by Koga *et al.* in 1981 [145]. It uses a uniform search pattern of nine locations (the center and endpoints of a $*$ shape). At each step, the step size is halved and the search pattern is centered at the minimum location from the previous step. The search is stopped when the step size is 1. The TSS starts with a step size of ± 4 pels in the first step, then ± 2 pels in the second step and ± 1 pel in the third step. This gives a maximum allowed displacement of $\pm 4 \pm 2 \pm 1 = \pm 7$ pels. For larger search windows the number of steps must be increased. This is called the N-steps search and is described in the following procedure.

1. Find the required number of steps N such that

$$2^{N-1} \leq d_m \leq 2^N.$$

2. Initialize the search step size to

$$s = 2^{N-1}.$$

3. Initialize the center of search to the origin of the search window:

$$(c_x, c_y) = (0, 0).$$

4. Evaluate the BDM at the center of the search and its eight neighbors at a step size of s. Out of this set of nine locations, find the one that achieves the minimum BDM:

$$(m_x, m_y) = \arg \min_{(i,j) \in \mathscr{P}} \text{BDM}(i, j),$$

where

$$\mathscr{P} = \{(c_x, c_y), (c_x + s, c_y), (c_x - s, c_y), (c_x, c_y + s), (c_x, c_y - s),$$
$$(c_x - s, c_y - s), (c_x - s, c_y + s), (c_x + s, c_y - s), (c_x + s, c_y + s)\}.$$

5. IF the search step size is 1, i.e., if $s = 1$, THEN

    (a) Set the final motion vector to the minimum location found so far:

$$(d_x, d_y) = (m_x, m_y).$$

    (b) STOP.

6. ELSE (when the step size is not 1, i.e. $s \neq 1$)

    (a) Halve the step size:

$$s = \frac{s}{2}.$$

    (b) Set the center of the search to the minimum location:

$$(c_x, c_y) = (m_x, m_y).$$

    (c) GOTO step 4.

A.4   The One-at-a-Time Search (OTS)

The one-at-a-time search (OTS) was proposed by Srinivasan and Rao in 1985 [146]. It uses two fixed-size uniform patterns. The search starts using a horizontal pattern of one center location and its immediate left and right neighbors. At each step, this search pattern is moved horizontally and centered at the minimum location from the previous step. This continues until the minimum is in the center of the pattern (i.e., the minimum is the same as that of the previous step). In this case, the search switches to the vertical direction using a pattern of one center location and its immediate top and bottom neighbors. This is explained in the following procedure.

1. Initialize the center of search to the origin of the search window:
$$(c_x, c_y) = (0, 0).$$

2. Evaluate the BDM at the center of the search and its immediate left and right neighbors. Out of this set of three locations, find the one that achieves the minimum BDM:

$$(m_x, m_y) = \arg \min_{(i,j) \in \mathscr{P}_h} \mathrm{BDM}(i, j),$$

where

$$\mathscr{P}_h = \{(c_x, c_y), (c_x + 1, c_y), (c_x - 1, c_y)\}.$$

3. IF the minimum is at the center of the search pattern, i.e., $(m_x, m_y) = (c_x, c_y)$, THEN GOTO step 5 (i.e., vertical direction).

4. ELSE

    (a) Move the center of the search to the new minimum location:

$$(c_x, c_y) = (m_x, m_y).$$

    (b) GOTO step 2 (i.e., continue in the horizontal direction).

5. Evaluate the BDM at the center of the search and its immediate top and bottom neighbors. Out of this set of three locations, find the one that achieves the minimum BDM:

$$(m_x, m_y) = \arg \min_{(i,j) \in \mathscr{P}_v} BDM(i, j),$$

where

$$\mathscr{P}_v = \{(c_x, c_y), (c_x, c_y + 1), (c_x, c_y - 1)\}.$$

6. IF the minimum is at the center of the search pattern, i.e., $(m_x, m_y) = (c_x, c_y)$, THEN

   (a) Set the final motion vector to the current minimum:

   $$(d_x, d_y) = (m_x, m_y)$$

   (b) STOP.

7. ELSE

   (a) Move the center of the search to the new minimum location:

   $$(c_x, c_y) = (m_x, m_y).$$

   (b) GOTO step 5 (i.e., continue in the vertical direction).

A.5   The Cross-Search Algorithm (CSA)

The cross-search algorithm (CSA) was proposed by Ghanbari in 1990 [147]. The search includes an early-termination criterion where, in the first step, a threshold is used to detect if the block is stationary. The search starts with a uniform pattern of five locations (the center and endpoints of an $\times$ shape). At each step, the search step size is halved and the search pattern is centered at the minimum location from the previous step. The search is stopped when the step size is 1. In this case, the search switches to one of two uniform patterns: five locations using either an $\times$ shape or a $+$ shape. This is explained in the following procedure.

1. Evaluate $BDM(0, 0)$.

2. IF $BDM(0, 0) <$ Threshold, THEN STOP.

3. Initialize the center of search to the origin of the search window:

$$(c_x, c_y) = (0, 0).$$

4. Initialize the search step size to half the maximum allowed displacement:

$$s = \left\lceil \frac{d_m}{2} \right\rceil.$$

5. Evaluate the BDM at the center of the search and its four diagonal neighbors at a step size of s. Out of this set of five locations, find the one that achieves the minimum BDM:

$$(m_x, m_y) = \arg \min_{(i,j) \in \mathscr{P}_1} \text{BDM}(i, j),$$

where

$$\mathscr{P}_1 = \{(c_x, c_y), (c_x - s, c_y - s), (c_x + s, c_y - s), (c_x - s, c_y + s), (c_x + s, c_y + s)\}.$$

6. IF search step size is 1, i.e., if $s = 1$, THEN

   (a) IF the minimum (m_x, m_y) is one of the three locations (c_x, c_y), $(c_x - 1, c_y - 1)$, or $(c_x + 1, c_y + 1)$, THEN

      i. Set the center of search to the minimum location:

      $$(c_x, c_y) = (m_x, m_y).$$

      ii. Evaluate the BDM at the center of the search and its four horizontal and vertical immediate neighbors. Out of this set of five locations, set the motion vector to the one that achieves the minimum BDM:

      $$(d_x, d_y) = \arg \min_{(i,j) \in \mathscr{P}_2} \text{BDM}(i, j),$$

      where

      $$\mathscr{P}_2 = \{(c_x, c_y), (c_x - 1, c_y), (c_x + 1, c_y), (c_x, c_y - 1), (c_x, c_y + 1)\}.$$

      iii. STOP.

   (b) ELSE

      i. Set the center of search to the minimum location:

      $$(c_x, c_y) = (m_x, m_y).$$

      ii. Evaluate the BDM at the center of the search and its four diagonal immediate neighbors. Out of this set of five locations, set the motion vector to the one that achieves the minimum BDM:

      $$(d_x, d_y) = \arg \min_{(i,j) \in \mathscr{P}_3} \text{BDM}(i, j),$$

      where

      $$\mathscr{P}_3 = \{(c_x, c_y), (c_x - 1, c_y - 1), (c_x + 1, c_y - 1),$$
      $$(c_x - 1, c_y + 1), (c_x + 1, c_y + 1)\}.$$

      iii. STOP.

7. ELSE (when step size is not 1, i.e., $s \neq 1$)

   (a) Halve the step size:
   $$s = \left\lceil \frac{s}{2} \right\rceil.$$

   (b) Set the center of the search to the minimum location:
   $$(c_x, c_y) = (m_x, m_y).$$

   (c) GOTO step 5.

A.6   The Diamond Search (DS)

The diamond search (DS) algorithm was proposed by Zhu and Ma in 1997 [150,151]. An identical version of the algorithm has also been proposed by Tham *et al.* in 1998 [149]. The algorithm uses two fixed-search patterns. It starts with a pattern of nine locations forming a diamond with a step size of 2. At each step, this search pattern is centered at the minimum location from the previous step. This process continues until the minimum is in the center of the pattern (i.e., the minimum is the same as that of the previous step). In this case, the algorithm switches to the second pattern. This consists of five locations forming a diamond with a step size of 1. This pattern is used only once and the search is then terminated. This is explained in the following procedure.

1. Initialize the center of search to the origin of the search window:

$$(c_x, c_y) = (0, 0).$$

2. Evaluate the BDM at nine locations forming a diamond with a step size of 2 centered at the current center location (c_x, c_y). Out of this set of nine locations, find the one that achieves the minimum BDM:

$$(m_x, m_y) = \arg \min_{(i,j) \in \mathscr{P}_{d_1}} BDM(i, j),$$

where

$$\mathscr{P}_{d_1} = \{(c_x, c_y), (c_x + 2, c_y), (c_x - 2, c_y), (c_x, c_y + 2), (c_x, c_y - 2)$$
$$(c_x - 1, c_y - 1), (c_x + 1, c_y - 1), (c_x - 1, c_y + 1), (c_x + 1, c_y + 1)\}.$$

3. IF the minimum is at the center of the search pattern, i.e., $(m_x, m_y) = (c_x, c_y)$, THEN

   (a) Evaluate the BDM at five locations forming a diamond with a step size of 1 centered at the current center location (c_x, c_y). Out of this set of five locations, set the motion vector to the one that achieves the minimum BDM:

   $$(d_x, d_y) = \arg \min_{(i,j) \in \mathscr{P}_{d_2}} BDM(i, j),$$

   where

   $$\mathscr{P}_{d_2} = \{(c_x, c_y), (c_x + 1, c_y), (c_x - 1, c_y), (c_x, c_y + 1), (c_x, c_y - 1)\}.$$

   (b) STOP.

4. ELSE (when the minimum is not in the center, i.e., $(m_x, m_y) \neq (c_x, c_y)$),

   (a) Set the center of the search to the new minimum location:

   $$(c_x, c_y) = (m_x, m_y).$$

   (b) GOTO step 2.

Bibliography

[1] UMTS Forum. The future mobile market: Global trends and developments with a focus on Western Europe. Report 8, UMTS Forum, March 1999. Download from http://www.umts-forum.org/reports.html.

[2] The GSM Association. http://www.gsmworld.com.

[3] Mobile GPRS. http://www.mobileGPRS.com.

[4] The Electronics & Communications Division of the IEE. Special issue on the universal mobile telecommunications system. *IEE Electronics & Communication Engineering Journal*, 12(3):89–152, June 2000.

[5] M. Budagavi, W. R. Heinzelman, J. Webb, and R. Talluri. Wireless MPEG-4 video communication on DSP chips. *IEEE Signal Processing Magazine*, 17(1):36–53, January 2000.

[6] A. Launiainen, A. Jore, E. Ryytty, T. Hämäläinen, and J. Saarinen. Evaluation of TMS320C62 performance in low-bit-rate video encoding. In *Proceedings of the IEEE Third Annual Multimedia and Applications Conference (MTAC)*, pages 364–368, Anaheim, CA, 15–17 September 1998.

[7] D. R. Bull, C. N. Canagarajah, and A. R. Nix, editors. *Insights into Mobile Multimedia Communications*. Signal Processing and Its Applications. Academic Press, London, 1999.

[8] D. T. Hoang, P. M. Long, and J. S. Vitter. Efficient cost measures for motion estimation at low bit rates. *IEEE Transactions on Circuits and Systems for Video Technology*, 8(4):488–500, August 1998.

[9] C. A. Poynton. Frequently asked questions about color. http://home.inforamp.net/~poynton/ColorFAQ.html.

[10] A. Murat Tekalp. *Digital Video Processing*. Prentice Hall Signal Processing Series. Prentice Hall, Englewood Cliffs, NJ, 1995.

[11] J. L. Mitchell, W. B. Pennebaker, C. E. Fogg, and D. J. LeGall. *MPEG Video Compression Standard*. Digital Multimedia Standards Series. Chapman & Hall, New York, 1996.

[12] C. A. Poynton. Frequently asked questions about gamma. http://home.inforamp.net/~poynton/GammaFAQ.html.

[13] A. N. Netravali and B. G. Haskell. *Digital Pictures: Representation, Compression and Standards*, 2nd edition. Applications of Communications Theory Series. Plenium Press, New York, 1995.

[14] CCIR. Recommendation 601-2: Encoding parameters of digital television for studios. In *Digital Methods of Transmitting Television Information*, pages 95–104. CCIR (currently ITU-R), 1990.

[15] M. Ghanbari. *Video Coding—an Introduction to Standard Codecs*. Volume 42 of IEE Telecommunications Series. The Institution of Electrical Engineers IEE, London, 1999.

[16] C. E. Shannon. A mathematical theory of communication. *Bell Systems Technical Journal*, 27(3):379–423, 1948.

[17] C. E. Shannon. Coding theorems for a discrete source with a fidelity criterion. Part 4, IRE National Convention Record, 1959.

[18] T. Berger. *Rate Distortion Theory*. Prentice-Hall, Englewood Cliffs, NJ, 1971.

[19] J. Max. Quantizing for minimum distortion. *IRE Transactions on Information Theory*, 6:7–12, March 1960.

[20] S. P. Lloyd. Least squares quantization in PCM. *IEEE Transactions on Information Theory*, 28:129–137, March 1982.

[21] R. C. Wood. On optimum quantization. *IEEE Transactions on Information Theory*, 15(2):248–252, 1969.

[22] D. A. Huffman. A method for the construction of minimum redundancy codes. *Proceedings of the IRE*, 40:1098–1101, 1952.

[23] M. Hankamer. A modified huffman procedure with reduced memory requirements. *IEEE Transactions on Communication*, 27(6):930–932, 1979.

[24] G. G. Langdon. An introduction to arithmetic coding. *IBM Journal, Research and Development*, 28(2):135–149, 1984.

[25] ITU-R. Methodology for the subjective assessment of the quality of television pictures. Recommendation BT.500-6, ITU-R, 1994.

[26] K. T. Tan, M. Ghanbari, and D. E. Pearson. An objective measurement tool for MPEG video quality. *Signal Processing*, 7:279–294, 1998.

[27] T. Alpert, V. Baroncini, D. Choi, L. Contin, R. Koenen, F. Pereira, and H. Peterson. Subjective evaluation of MPEG-4 video codec proposals: Methodological approach and test procedures. *Signal Processing: Image Communication*, 9(4):305–325, May 1997.

[28] C. C. Cutler. Differential quantization of communication signals. U.S. Patent No. 2605361, 29 July 1952.

[29] M. Rabbani and P. W. Jones. *Digital Image Compression Techniques*. Volume TT7 of SPIE Tutorial Texts in Optical Engineering. SPIE–The International Society for Optical Engineering, Washington, DC, 1991.

[30] R. J. Clarke. *Transform Coding of Images*. Microelectronics and Signal Processing. Academic Press, London, 1985.

[31] E. Feig and S. Winograd. Fast algorithms for the discrete cosine transform. *IEEE Transactions on Signal Processing*, 40(9):2174–2193, September 1992.

[32] N. Ahmed, T. Natarajan, and K. R. Rao. Discrete cosine transform. *IEEE Transactions on Computers*, C-23:90–93, January 1974.

[33] K. R. Rao and P. Yip. *Discrete Cosine Transform—Algorithms, Advantages, Applications*. Academic Press, San Diego, CA, 1990.

[34] D. E. Pearson and M. W. Whybray. Transform coding of images using interleaved blocks. *IEE Proceedings, Part F*, 131:466–472, August 1984.

[35] H. S. Malvar and D. H. Staelin. The LOT: Transform coding without blocking effects. *IEEE Transactions on Acoustics, Speech and Signal Processing*, 37(4):553–559, April 1989.

[36] S. Minami and A. Zakhor. An optimization approach for removing blocking effects in transform coding. *IEEE Transactions on Circuits and Systems for Video Technology*, 5(2):74–82, April 1995.

[37] S. Nanda and W. A. Pearlman. Tree coding of image subbands. *IEEE Transactions on Image Processing*, 1(2):133–147, 1992.

[38] R. E. Crochiere, S. A. Webber, and F. L. Flanagan. Digital coding of speech in subbands. *Bell Systems Technical Journal*, 55(8):1069–1085, 1976.

[39] J. Woods and S. O'Neil. Subband coding of images. *IEEE Transactions on Acoustics, Speech and Signal Processing*, ASSP-34:1278–1288, October 1986.

[40] J. M. Shapiro. Embedded image coding using zerotrees of wavelet coefficients. *IEEE Transactions on Signal Processing*, 41:3445–3462, 1993.

[41] A. Said and W. A. Pearlman. Image compression using the spatial-orientation tree. In *Proceedings of the IEEE International Symposium on Circuits and Systems (ISCAS)*, pages 279–282, Chicago, May 1993.

[42] A. Said and W. A. Pearlman. A new fast and efficient image codec based on set partitioning in hierarchical trees. *IEEE Transactions on Circuits and Systems for Video Technology*, 6:243–250, June 1996.

[43] O. Egger, W. Li, and M. Kunt. High compression image coding using an adaptive morphological subband decomposition. *Proceedings of the IEEE*, 83:272–287, February 1995.

[44] D. W. Redmill and D. R. Bull. Non-linear perfect reconstruction filter banks for image coding. In *Proceedings of the IEEE International*

Conference on Image Processing (ICIP), pages 593–596, Lausanne, Switzerland, 16–19 September 1996.

[45] Y. Linde, A. Buzo, and R. M. Gray. An algorithm for vector quantizer design. *IEEE Transactions on Communications*, COM-28(1):84–95, 1980.

[46] P. A. Chou, T. Lookabaugh, and R. M. Gray. Entropy-constrained vector quantization. *IEEE Transactions on Acoustics, Speech and Signal Processing*, 37(1):31–42, January 1989.

[47] Chang-Hsing Lee and Ling-Hwei Chen. A fast search algorithm for vector quantization using mean pyramids of codewords. *IEEE Transactions on Communications*, 43:1697–1702, 1995.

[48] T. D. Lookabaugh and R. M. Gray. High-resolution quantization theory and the vector quantizer advantage. *IEEE Transactions on Information Theory*, IT-35:1020–1033, 1989.

[49] M. Kunt, A. Ikonomopoulos, and M. Kocher. Second-generation image-coding techniques. *Proceedings of the IEEE*, 73(4):549–574, April 1985.

[50] M. Kunt, M. Bernard, and R. Leonardi. Recent results in high-compression image coding. *IEEE Transactions on Circuits and Systems*, CAS-34(11):1306–1336, November 1987.

[51] M. E. Al-Mualla. Second-generation image coding techniques. Master's thesis, University of Bristol, Faculty of Engineering, Department of Electrical and Electronics Engineering, October 1996.

[52] R. J. Clarke. *Digital Compression of Still Images and Video*. Signal Processing and Its Applications. Academic Press, London, 1995.

[53] B. G. Haskell, F. W. Mounts, and J. C. Candy. Interframe coding of videotelephone pictures. *Proceedings of the IEEE*, 60:792–800, July 1972.

[54] J. R. Jain and A. K. Jain. Displacement measurement and its application in interframe image coding. *IEEE Transactions on Communications*, COM-29(12):1799–1808, December 1981.

[55] F. Dufaux and F. Moscheni. Motion estimation techniques for digital TV: A review and a new contribution. *Proceedings of the IEEE*, 83(6): 858–875, June 1995.

[56] M. F. Chowdhury, A. F. Clark, A. C. Downton, and D. E. Pearson. A switched model-based coder for video signals. *IEEE Transactions on Circuits and Systems for Video Technology*, 4:216–217, June 1994.

[57] D. E. Pearson. Developments in model-based video coding. *Proceedings of the IEEE*, 83(6):892–906, June 1995.

[58] CCITT/SG XV. Codecs for videoconferencing using primary digital group transmission. Recommendation H.120, CCITT (currently ITU-T), Geneva, 1989.

[59] CCITT/SG XV. Video codec for audiovisual services at $p \times 64$ kbit/s. Recommendation H.261, CCITT (currently ITU- T), Geneva, 1993.

[60] CCIR. Transmission of component-coded digital television signals for contribution-quality applications at bit rates near 140 Mbit/s. Recommendation 721, CCIR (currently ITU-R), Geneva, 1990.

[61] CCIR. Digital coding of component television signals for contribution-quality applications in the range 34–45 Mbit/s. Recommendation 723, CCIR (currently ITU-R), Geneva, 1992.

[62] ISO/IEC JTC1/SC29/WG11. Information technology—Coding of moving pictures and associated audio for digital storage media at up to about 1.5 Mbits/s. Part 2: Video. Draft ISO/IEC 11172-2 (MPEG-1), ISO/IEC, Geneva, 1991.

[63] ISO/IEC JTC1/SC29/WG11 and ITU-T/SG15. Information technology—Generic coding of moving pictures and associated audio. Part 2: Video. Draft ISO/IEC 13818-2 (MPEG-2) and ITU-T Recommendation H.262, ISO/IEC and ITU-T, Geneva, 1994.

[64] ITU-T/SG15. Video coding for low bitrate communication. ITU-T Recommendation H.263, Version 1, ITU-T, Geneva, 1996.

[65] B. Girod, E. Steinbach, and N. Färber. Performance of the H.263 video compression standard. *Journal of VLSI Signal Processing: Systems for Signal, Image, and Video Technology*, 17:101–111, 1997.

[66] ITU-T/SG16/Q15. Video coding for low bitrate communication. ITU-T Recommendation H.263, Version 2 (H.263+), ITU-T, Geneva, 1998.

[67] ISO/IEC JTC1/SC29/WG11. Information technology—Generic coding of audio-visual objects. Part 2: Visual. Draft ISO/IEC 14496-2 (MPEG-4), Version 1, ISO/IEC, Geneva, 1998.

[68] ITU-T/SG16/Q15. Draft for "H.263++" annexes U, V, and W to recommendation H.263. Draft, ITU-T, Geneva, 2000.

[69] R. Schäfer and T. Sikora. Digital video coding standards and their role in video communications. *Proceedings of the IEEE*, 83(6):907–924, June 1995.

[70] T. Ebrahimi and M. Kunt. Visual data compression for multimedia applications. *Proceedings of the IEEE*, 86(6):1109–1125, June 1998.

[71] G. Côté, B. Erol, M. Gallant, and F. Kossentini. H.263+: Video coding at low bit rates. *IEEE Transactions on Circuits and Systems for Video Technology*, 8(7):849–866, November 1998.

[72] S. Wenger, G. Knorr, J. Ott, and F. Kossentini. Error resilience support in H.263+. *IEEE Transactions on Circuits and Systems for Video Technology*, 8(7):867–877, November 1998.

[73] T. Sikora. MPEG digital video-coding standards. *IEEE Signal Processing Magazine*, pages 82–100, September 1997.

[74] F. Pereira, K. O'Connell, R. Koenen, and M. Etoh. Special issue on MPEG-4, part 1: Invited papers. *Signal Processing: Image Communication*, 9(4):291–477, May 1997.

[75] F. Pereira. Tutorial issue on the MPEG-4 standard. *Signal Processing: Image Communication*, 15(4–5):269–478, January 2000.

[76] Telenor Research. Video codec test model. TMN5, ITU-T/SG15/WP 15/1 Expert's Group on Very Low Bitrate Visual Telephony, Geneva, 1995.

[77] Y. T. Tse and R. L. Baker. Global zoom/pan estimation and compensation for video compression. In *Proceedings of the IEEE International Conference on Acoustics, Speech and Signal Processing (ICASSP)*, volume 4, pages 2725–2728, Toronto, May 1991.

[78] C. R. Moloney and E. Dubois. Estimation of motion fields from image sequences with illumination variation. In *Proceedings of the IEEE International Conference on Acoustics, Speech and Signal Processing (ICASSP)*, volume 4, pages 2425–2428, Toronto, May 1991.

[79] M. Bertero, T. A. Poggio, and V. Torre. Ill-posed problems in early vision. *Proceedings of the IEEE*, 76(8):869–889, August 1988.

[80] C. Stiller and J. Konrad. Estimating motion in image sequences. *IEEE Signal Processing Magazine*, 16(4):70–91, July 1999.

[81] J. O. Limb and J. A. Murphy. Measuring the speed of moving objects from television signals. *IEEE Transactions on Communication*, COM-23(4):474–478, April 1975.

[82] C. Cafforio and F. Rocca. Methods for measuring small displacements of television images. *IEEE Transactions on Information Theory*, IT-22(5):573–579, September 1976.

[83] H. Yamaguchi. Iterative method of movement estimation for television signals. *IEEE Transactions on Communications*, 37(12):1350–1358, December 1989.

[84] H. M. Ming, Y. M. Chou, and S. C. Cheng. Motion estimation for video coding standards. *Journal of VLSI Signal Processing: Systems for Signal, Image, and Video Technology*, 17:113–136, 1997.

[85] A. N. Netravali and J. D. Robbins. Motion compensated television coding: Part I. *Bell System Technical Journal*, 58:631–670, March 1979.

[86] D. R. Walker and K. R. Rao. Improved pel-recursive motion compensation. *IEEE Transactions on Communications*, COM-32(10):1128–1134, October 1984.

[87] H. G. Musmann, P. Pirsch, and H. J. Grallert. Advances in picture coding. *Proceedings of the IEEE*, 73(4):523–548, April 1985.

[88] B. G. Haskell. Frame-to-frame coding of television pictures using two-dimensional Fourier transforms. *IEEE Transactions on Information Theory*, 20:119–120, January 1974.

[89] P. A. Lynn and W. Fuerst. *Introductory Digital Signal Processing with Computer Applications*, revised edition, Wiley, London, 1994.

[90] C. D. Kuglin and D. C. Hines. The phase correlation image alignment method. In *Proceedings of the IEEE International Conference on Cybernetics and Society*, pages 163–165, San Francisco, 1975.

[91] G. A. Thomas. Television motion measurement for DATV and other applications. Technical Report 1987/11, British Broadcasting Corporation (BBC) Research Department, 1987.

[92] B. Girod. Motion-compensating prediction with fractional-pel accuracy. *IEEE Transactions on Communications*, 41(4):604–612, April 1993.

[93] Snell & Wilcox Ltd. Alchemist Ph.C D: a phase correlation 10-bit motion compensated standards converter for digital I/O. http://www.snellwilcox.com/productguide/linker1/alc2index.html.

[94] Y. M. Chou and H. M. Hang. A new motion estimation method using frequency components. *Journal of Visual Communication and Image Representation*, 8(1):83–96, March 1997.

[95] R. W. Young and N. G. Kingsbury. Frequency-domain motion estimation using a complex lapped transform. *IEEE Transactions on Image Processing*, 2:2–17, January 1993.

[96] U. V. Koc and K. J. R. Liu. DCT-based motion estimation. *IEEE Transactions on Image Processing*, 7(7):948–965, July 1998.

[97] U. V. Koc and K. J. R. Liu. Interpolation-free subpixel motion estimation techniques in DCT domain. *IEEE Transactions on Circuits and Systems for Video Technology*, 8(4):460–487, August 1998.

[98] M. H. Chan, Y. B. Yu, and A. G. Constantinides. Variable size block matching motion compensation with applications to video coding. *IEE Proceedings, Part I*, 137(4):205–212, August 1990.

[99] G. J. Sullivan and R. L. Baker. Efficient quadtree coding of images and video. In *Proceedings of the IEEE International Conference on Acoustics, Speech and Signal Processing (ICASSP)*, pages 2661–2664, Toronto, May 1991.

[100] S. Ericsson. Fixed and adaptive predictors for hybrid predictive/transform coding. *IEEE Transactions on Communications*, COM-33(12):1291–1302, December 1985.

[101] H. Watanabe and S. Singhal. Windowed motion compensation. In *Proceedings of the SPIE Conference on Visual Communications and Image Processing (VCIP)*, volume 1605, pages 582–589, November 1991.

[102] C. Auyeung, J. Kosmach, M. Orchard, and T. Kalafatis. Overlapped block motion compensation. In *Proceedings of the SPIE Conference on Visual Communications and Image Processing (VCIP)*, volume 1818, pages 561–572, November 1992.

[103] M. T. Orchard and G. J. Sullivan. Overlapped block motion compensation: An estimation-theoretic approach. *IEEE Transactions on Image Processing*, 3(5):693–699, September 1994.

[104] T. Y. Kuo and C. C. J. Kuo. Fast overlapped block motion compensation with checkerboard block partitioning. *IEEE Transactions on Circuits and Systems for Video Technology*, 8(6):705–712, October 1998.

[105] G. J. Sullivan. Multi-hypothesis motion compensation for low-bit-rate video coding. In *Proceedings of the IEEE International Conference on Acoustics, Speech and Signal Processing (ICASSP)*, volume 5, pages 437–440, Minneapolis, April 1993.

[106] B. Girod. Efficiency analysis of multihypothesis motion-compensated prediction for video coding. *IEEE Transactions on Image Processing*, 9(2):173–183, February 2000.

[107] G. Wolberg. *Digital Image Warping*. IEEE Computer Society Press, Los Alamitos, CA, 1990.

[108] G. J. Sullivan and R. L. Baker. Motion compensation for video compression using control grid interpolation. In *Proceedings of the IEEE International Conference on Acoustics, Speech and Signal Processing (ICASSP)*, volume 4, pages 2713–2716, Toronto, May 1991.

[109] C. L. Huang and C. Y. Hsu. A new motion compensation method for image sequence coding using hierarchical grid interpolation. *IEEE Transactions on Circuits and Systems for Video Technology*, 4(1): 42–51, February 1994.

[110] J. Niewęgłowski and P. Haavisto. Temporal image sequence prediction using motion field interpolation. *Signal Processing: Image Communication*, 7:333–353, 1995.

[111] J. Niewęgłowski, T. G. Campbell, and P. Haavisto. A novel video coding scheme based on temporal prediction using digital image warping. *IEEE Transactions on Consumer Electronics*, 39(3):141–150, August 1993.

[112] A. Neri and S. Colonnese. On the computation of warping-based motion compensation in video sequence coding. *Signal Processing: Image Communication*, 13:155–160, 1998.

[113] A. Nosratinia and M. T. Orchard. Optimal unified approach to warping and overlapped block motion estimation in video coding. In *Proceedings of the SPIE Conference on Visual Communications and Image Processing (VCIP)*, volume 2727, pages 634–644, Orlando, FL, March 1996.

[114] Y. Nakaya and H. Harashima. Motion compensation based on spatial transformations. *IEEE Transactions on Circuits and Systems for Video Technology*, 4(3):339–356, June 1994.

[115] M. Ghanbari, S. de Faria, I. N. Goh, and K. T. Tan. Motion compensation for very low-bit-rate video. *Signal Processing: Image Communication*, 7:567–580, 1995.

[116] A. Sharaf and F. Marvasti. Motion compensation using spatial transformations with forward mapping. *Signal Processing: Image Communication*, 14:209–227, 1999.

[117] F. J. P. Lopes and M. Ghanbari. Analysis of spatial transform motion estimation with overlapped compensation and fractional-pixel accuracy. *IEE Proceedings on Vision, Image and Signal Processing*, 146(6):339–344, December 1999.

[118] C. A. Papadopoulos and T. G. Clarkson. Motion compensation using second-order geometric transformations. *IEEE Transactions on Circuits and Systems for Video Technology*, 5(4):319–331, August 1995.

[119] V. Seferidis and M. Ghanbari. General approach to block-matching motion estimation. *Optical Engineering*, 32(7):1464–1474, July 1993.

[120] V. Seferidis and M. Ghanbari. Generalised block-matching motion estimation using quad-tree structured spatial decomposition. *IEE Proceedings on Vision, Image and Signal Processing*, 141(6):446–452, December 1994.

[121] Y. Wang and O. Lee. Active mesh—a feature seeking and tracking image sequence representation scheme. *IEEE Transactions on Image Processing*, 3(5):610–624, September 1994.

[122] Y. Wang and O. Lee. Use of two-dimensional deformable mesh structures for video coding. Part I—the synthesis problem: Mesh-based function approximation and mapping. *IEEE Transactions on Circuits and Systems for Video Technology*, 6(6):637–646, December 1996.

[123] Y. Wang, O. Lee, and A. Vetro. Use of two-dimensional deformable mesh structures for video coding. Part II—the analysis problem and a region-based coder employing an active mesh representation. *IEEE Transactions on Circuits and Systems for Video Technology*, 6(6):647–659, December 1996.

[124] M. Dudon, O. Avaro, and C. Roux. Triangular active mesh for motion estimation. *Signal Processing: Image Communication*, 10:21–41, 1997.

[125] Y. Altunbasak and A. M. Tekalp. Closed-form connectivity-preserving solutions for motion compensation using 2-D meshes. *IEEE Transactions on Image Processing*, 6(9):1255–1269, September 1997.

[126] Y. Wang and J. Osterman. Evaluation of mesh-based motion estimation in H.263-like coders. *IEEE Transactions on Circuits and Systems for Video Technology*, 8(3):243–252, June 1998.

[127] A. M. Tekalp, P. V. Beek, C. Toklu, and B. Günsel. Two-dimensional mesh-based visual-object representation for interactive synthetic/natural digital video. *Proceedings of the IEEE*, 86(6):1029–1051, June 1998.

[128] H. Brusewitz. Motion compensation with triangles. In *Proceedings of the 3rd International Workshop on 64 kbits/s Coding of Moving Video*, Free session, Rotterdam, September 1990.

[129] K. T. Tan, I. N. Goh, and M. Ghanbari. Fast motion estimation with spatial transformation. *IEE Electronics Letters*, 30:847–849, 1994.

[130] D. B. Bradshaw and N. G. Kingsbury. A fast, two-stage translational and warping motion compensation scheme. In *Proceedings of the IX European Signal Processing Conference (EUSIPCO)*, volume II, pages 905–908, Island of Rhodes, Greece, 1998.

[131] ISO/IEC JTC1/SC29/WG11. Core experiment on global motion compensation (P1). In *Description of Core Experiments on Coding Efficiency in MPEG-4 Video*, No. N1385, September 1996.

[132] ISO/IEC JTC1/SC29/WG11. Core experiment N3: Dynamic sprites and global motion compensation. In *Description of Core Experiments on Coding Efficiency in MPEG-4 Video*, No. N1875, October 1997.

[133] ISO/IEC JTC1/SC29/WG11. Core experiments on STFM/LTFM for motion prediction (P3). In *Description of Core Experiments on Coding Efficiency in MPEG-4 Video*, No. N1385, September 1996.

[134] N. Mukawa and H. Kuroda. Uncovered background prediction in interframe coding. *IEEE Transactions on Communications*, COM-33(11):1227–1231, November 1985.

[135] T. Wiegand, X. Zhang, and B. Girod. Motion-compensating long-term memory prediction. In *Proceedings of the IEEE International Conference on Image Processing (ICIP)*, volume 2, pages 53–56, Santa Barbara, CA, 26–29 October 1997.

[136] T. Wiegand, X. Zhang, and B. Girod. Long-term memory motion-compensated prediction. *IEEE Transactions on Circuits and Systems for Video Technology*, 9(1):70–84, February 1999.

[137] T. Wiegand, E. Steinbach, A. Stensrud, and B. Girod. Multiple-reference picture video coding using polynomial motion models. In *Proceedings of the SPIE Conference on Visual Communications and Image Processing (VCIP)*, volume 3309, pages 134–145, San Jose, CA, February 1998.

[138] E. Steinbach, T. Wiegand, and B. Girod. Using multiple global models for improved block-based video coding. In *Proceedings of the IEEE International Conference on Image Processing (ICIP)*, volume 2, pages 56–60, Kobe, Japan, October 1999.

[139] T. Wiegand, E. Steinbach, and B. Girod. Long-term memory prediction using affine motion compensation. In *Proceedings of the IEEE International Conference on Image Processing (ICIP)*, volume 1, pages 51–55, Kobe, Japan, October 1999.

[140] T. Wiegand, N. Färber, B. Girod, and B. Andrews. Proposed draft for annex on enhanced reference picture selection. Document Q15-F-32r1, ITU-T/SG16/Q.15, Seoul, Korea, November 1998. Downnlond from http://www-nt.etechnik.uni-erlangen.de/~wiegand/publications.html.

[141] T. Wiegand, B. Lincoln, and B. Girod. Fast search for long-term memory motion-compensated prediction. In *Proceedings of the IEEE International Conference on Image Processing (ICIP)*, volume 3, pages 619–622, Chicago, 4–7 October 1998.

[142] T. R. Gardos, editor. Video codec test model, near-term, version 10 (TMN10) draft 1. Document Q15-D-65d1, ITU-T/SG16/Q.15, Tampere, Finland, April 1998. Download from ftp://standard.pictel.com/video-site/h263plus/tmn10.doc.

[143] G. J. Sullivan and T. Wiegand. Rate-distortion optimization for video compression. *IEEE Signal Processing Magazine*, 15(6):74–90, November 1998.

[144] Telenor Research and Development. H.263 TMN-software codec, version 2.0. ftp://bonde.nta.no/pub/tmn/software.

[145] T. Koga, K. Iinuma, A. Hirano, Y. Iijima, and T. Ishiguro. Motion-compensated interframe coding for video conferencing. In *Proceedings of the National Telecommunications Conference (NTC)*, pages G5.3.1–G5.3.5, New Orleans, November 29–December 3 1981.

[146] R. Srinivasan and K. R. Rao. Predictive coding based on efficient motion estimation. *IEEE Transactions on Communications*, COM-33(8):888–896, August 1985.

[147] M. Ghanbari. The cross-search algorithm. *IEEE Transactions on Communications*, 38(7):950–953, July 1990.

[148] K. H. K. Chow and M. L. Liou. Genetic motion search algorithm for video compression. *IEEE Transactions on Circuits and Systems for Video Technology*, 3(6):440–445, December 1993.

[149] J. Y. Tham, S. Ranganath, M. Ranganath, and A. A. Kassim. A novel unrestricted center-biased diamond search algorithm for block motion estimation. *IEEE Transactions on Circuits and Systems for Video Technology*, 8(4):369–377, August 1998.

[150] S. Zhu and K. K. Ma. A new diamond search algorithm for fast block matching motion estimation. In *Proceedings of the International Conference on Information, Communication and Signal Processing (ICICS)*, pages 292–296, 9–12 September 1997.

[151] S. Zhu and K. K. Ma. A new diamond search algorithm for fast block-matching motion estimation. *IEEE Transactions on Image Processing*, 9(2):287–290, February 2000.

[152] H. Ghavari and M. Mills. Blockmatching motion estimation—New results. *IEEE Transactions on Circuits and Systems*, 37(5):649–651, May 1990.

[153] M. J. Chen, L. G. Chen, T. D. Chiueh, and Y. P. Lee. A new block-matching criterion for motion estimation and its implementation. *IEEE Transactions on Circuits and Systems for Video Technology*, 5(3):231–236, June 1995.

[154] Y. Baek, H. S. Oh, and H. K. Lee. An efficient block-matching criterion for motion estimation and its VLSI implementation. *IEEE Transactions on Consumer Electronics*, 42(4):885–892, November 1996.

[155] K. Sauer and B. Schwartz. Efficient block motion estimation using integral projections. *IEEE Transactions on Circuits and Systems for Video Technology*, 6(5):513–518, October 1996.

[156] B. Natarajan, V. Bhaskaran, and K. Konstantinides. Low-complexity block-based motion estimation via one-bit transforms. *IEEE Transactions on Circuits and Systems for Video Technology*, 7(4):702–706, August 1997.

[157] B. Liu and A. Zaccarin. New fast algorithms for the estimation of block motion vectors. *IEEE Transactions on Circuits and Systems for Video Technology*, 3(2):148–157, April 1993.

[158] Y. L. Chan and W. C. Siu. New adaptive pixel decimation for block motion vector estimation. *IEEE Transactions on Circuits and Systems for Video Technology*, 6(1):113–118, February 1996.

[159] M. Bierling. Displacement estimation by hierarchical blockmatching. In *Proceedings of the SPIE Visual Communications and Image Processing (VCIP)*, volume 1001, pages 942–951, 1988.

[160] K. M. Nam, J. S. Kim, R. H. Park, and Y. S. Shim. A fast hierarchical motion vector estimation algorithm using mean pyramid. *IEEE Transactions on Circuits and Systems for Video Technology*, 5(4):344–351, August 1995.

[161] C. D. Bei and R. M. Gray. An improvement of the minimum distortion encoding algorithm for vector quantization. *IEEE Transactions on Communications*, COM-33(10):1132–1133, October 1985.

[162] W. Li and E. Salari. Successive elimination algorithm for motion estimation. *IEEE Transactions on Image Processing*, 4(1):105–107, January 1995.

[163] S. H. Huang and S. H. Chen. Fast encoding algorithm for VQ-based image coding. *IEE Electronics Letters*, 26(19):1618–1619, 13 September 1990.

[164] Y. C. Lin and S. C. Tai. Fast full-search block-matching algorithm for motion-compensated video compression. *IEEE Transactions on Communications*, 45(5):527–531, May 1997.

[165] M. E. Al-Mualla, C. N. Canagarajah, and D. R. Bull. A fast block matching motion estimation algorithm based on simplex minimisation. In *Proceedings of the IX European Signal Processing Conference (EUSIPCO)*, volume III, pages 1565–1568, Island of Rhodes, Greece, 8–11 September 1998.

[166] M. E. Al-Mualla, C. N. Canagarajah, and D. R. Bull. Simplex minimisation for fast block matching motion estimation. *IEE Electronics Letters*, 34(4):351–352, 19 February 1998.

[167] M. E. Al-Mualla, C. N. Canagarajah, and D. R. Bull. Simplex minimisation for multiple-reference motion estimation. In *Proceedings of the IEEE International Symposium on Circuits and Systems (ISCAS)*, volume IV, pages 733–736, Geneva, 28–31 May 2000.

[168] M. E. Al-Mualla, C. N. Canagarajah, and D. R. Bull. Simplex minimization for fast long-term memory motion estimation. *IEE Electronics Letters*, 37(5):290–292, 1 March 2001.

[169] M. E. Al-Mualla, C. N. Canagarajah, and D. R. Bull. Simplex minimization for single- and multiple-reference motion estimation. *IEEE Transactions on Circuits and Systems for Video Technology*, 11(12):1209–1220, December 2001.

[170] D. E. Knuth. Searching and sorting. In *The Art of Computer Programming*, volume 3. Addison-Wesley, Reading, MA, 1973.

[171] M. R. Hestenes. *Conjugate Direction Methods in Optimization*. Springer-Verlag, New York, 1980.

[172] D. E. Goldberg. *Genetic Algorithms in Search, Optimization and Machine Learning*. Addison-Wesley, Reading, MA, 1989.

[173] J. A. Nelder and R. Mead. A simplex method for function minimization. *The Computer Journal*, 7:308–313, 1965.

[174] W. H. Press, S. A. Teukolsky, W. T. Vetterling, and B. P. Flannery. *Numerical Recipes in C: The Art of Scientific Computing*, 2nd ed. Cambridge University Press, New York, 1992.

[175] O. Sohm. Fast block motion estimation algorithm for MPEG-4. Progress Report for the Project: MPEG-4 Video Coding Using the TMS320C62x, Image Communications Group, Center for Communications Research, University of Bristol, U.K., March 2000.

[176] ISO/IEC JTC1/SC29/WG11. MPEG-4 video verification model, version 14.0. Document N2932, ISO/IEC, October 1999.

[177] D. W. Redmill. *Image and Video Coding for Noisy Channels*. PhD thesis, University of Cambridge, Department of Engineering, Signal Processing and Communications Laboratory, November 1994.

[178] T. J. Ferguson and J. H. Rabinowitz. Self-synchronizing Huffman codes. *IEEE Transactions on Information Theory*, 30:687–693, 1984.

[179] Y. Wang and Q. F. Zhu. Error control and concealment for video communication: A review. *Proceedings of the IEEE*, 86(5):974–997, May 1998.

[180] Y. Wang, S. Wenger, J. Wen, and A. K. Kastaggelos. Error resilient video coding techniques. *IEEE Signal Processing Magazine*, 17(4):61–82, July 2000.

[181] P. Haskell and D. Messerschmitt. Resynchronization of motion compensated video affected by ATM cell loss. In *Proceedings of the IEEE International Conference on Acoustics, Speech and Signal Processing (ICASSP)*, volume III, pages 545–548, San Francisco, March 1992.

[182] G. Côté and F. Kossentini. Optimal intracoding of blocks for robust video communication over the internet. *Signal Processing: Image Communication*, 15(1–2):25–34, September 1999.

[183] J. Y. Liao and J. D. Villasenor. Adaptive intra update for video coding over noisy channels. In *Proceedings of the IEEE International Conference on Image Processing (ICIP)*, volume III, pages 763–766, Lausanne, Switzerland, 16–19 September 1996.

[184] G. Côté, S. Shirani, and F. Kossentini. Optimal mode selection and synchronization for robust video communications over error-prone networks. *IEEE Journal on Selected Areas in Communications*, 18(6):952–965, June 2000.

[185] D. W. Redmill and N. G. Kingsbury. The EREC: An error-resilient technique for coding variable-length blocks of data. *IEEE Transactions on Image Processing*, 5(4):565–574, April 1996.

[186] M. Ghanbari. Two-layer coding of video signals for VBR networks. *IEEE Journal on Selected Areas in Communications*, 7(5):771–781, June 1989.

[187] M. Khansari and M. Vetterli. Layered transmission of signals over power-constrained wireless channels. In *Proceedings of the IEEE International Conference on Image Processing (ICIP)*, volume III, pages 380–383, Washington, DC, October 1995.

[188] V. A. Vaishampayan. Design of multiple description scalar quantizers. *IEEE Transactions on Information Theory*, 39(3):821–834, May 1993.

[189] Q. F. Zhu, Y. Wang, and L. Shaw. Coding and cell-loss recovery in DCT-based packet video. *IEEE Transactions on Circuits and Systems for Video Technology*, 3(3):248–258, June 1993.

[190] M. Ghanbari and V. Seferidis. Cell-loss concealment in ATM video codecs. *IEEE Transactions on Circuits and Systems for Video Technology*, 3(3):238–247, June 1993.

[191] P. Salama, N. B. Shroff, E. J. Coyle, and E. J. Delp. Error concealment techniques for encoded video streams. In *Proceedings of the IEEE*

International Conference on Image Processing (ICIP), volume I, pages 9–12, Washington, DC, 23–26 October 1995.

[192] A. Narula and J. Lim. Error concealment techniques for an all-digital high-definition television system. In *Proceedings of the SPIE Conference on Visual Communications and Image Processing (VCIP)*, volume 2094, pages 304–315, 1993.

[193] Y. W. Wang, Q. F. Zhu, and L. Shaw. Maximally smooth image recovery in transform coding. *IEEE Transactions on Communications*, 41(10):1544–1551, October 1993.

[194] W. Kwok and H. Sun. Multi-directional interpolation for spatial error concealment. *IEEE Transactions on Consumer Electronics*, 39(3): 455–460, August 1993.

[195] P. Salama, N. B. Shroff, and E. J. Delp. A Bayesian approach to error concealment in encoded video streams. In *Proceedings of the IEEE International Conference on Image Processing (ICIP)*, volume II, pages 49–51, Lausanne, Switzerland, 16–19 September 1996.

[196] W. M. Lam, A. R. Reibman, and B. Liu. Recovery of lost or erroneously received motion vectors. In *Proceedings of the IEEE International Conference on Acoustic, Speech and Signal Processing ICASSP*, volume V, pages 417–420, Minnesota, U.S.A., April 1993.

[197] K. W. Kang, S. H. Lee, and T. Kim. Recovery of coded video sequences from channel errors. In *Proceedings of the SPIE Conference on Visual Communications and Image Processing (VCIP)*, volume 2501, pages 19–27, 1995.

[198] M. J. Chen, L. G. Chen, and R. M. Weng. Error concealment of lost motion vectors with overlapped motion compensation. *IEEE Transactions on Circuits and Systems for Video Technology*, 7(3):560–563, June 1997.

[199] S. Shirani, F. Kossentini, and R. Ward. A concealment method for video communications in an error-prone environment. *IEEE Transactions on Selected Areas in Communications*, 18(6):1122–1128, June 2000.

[200] H. Sun, K. Challapali, and J. Zdepski. Error concealment in digital simulcast AD-HDTV decoder. *IEEE Transactions on Consumer Electronics*, 38(3):108–117, August 1992.

[201] B. Girod and N. Färber. Feedback-based error control for mobile video transmission. *Proceedings of the IEEE*, 87(10):1707–1723, October 1999.

[202] E. Steinbach, N. Färber, and N. Girod. Standard compatible extension of H.263 for robust video transmission in mobile environments. *IEEE Transactions on Circuits and Systems for Video Technology*, 7(6):872–881, December 1997.

[203]   M. Wada. Selective recovery of video packet loss using error concealment. *IEEE Journal on Selected Areas in Communications*, 7(5):807–814, June 1989.

[204]   M. E. Al-Mualla, C. N. Canagarajah, and D. R. Bull. Error concealment using motion field interpolation. In *Proceedings of the IEEE International Conference on Image Processing (ICIP)*, volume II, pages 512–516, Chicago, 4–7 October 1998.

[205]   M. E. Al-Mualla, C. N. Canagarajah, and D. R. Bull. Temporal error concealment using motion field interpolation. *IEE Electronics Letters*, 35(3):215–217, 4 February 1999.

[206]   M. E. Al-Mualla, C. N. Canagarajah, and D. R. Bull. On the performance of temporal error concealment for long-term motion-compensated prediction. In *Proceedings of the IEEE International Conference on Image Processing (ICIP)*, volume III, pages 376–379, Vancouver, 10–13 September 2000.

[207]   M. E. Al-Mualla, C. N. Canagarajah, and D. R. Bull. Motion field interpolation for temporal error concealment. *IEE-Proceedings, Vision, Image and Signal Processing*, 147(5):445–453, October 2000.

[208]   M. E. Al-Mualla, C. N. Canagarajah, and D. R. Bull. Multiple-reference temporal error concealment. In *Proceedings of the IEEE International Symposium on Circuits and Systems (ISCAS)*, volume V, pages 149–152, Sydney, 6–9 May 2001.

[209]   A. Nosratinia. New kernels for fast mesh-based motion estimation. *IEEE Transactions on Circuits and Systems for Video Technology*, 11(1):40–51, January 2001.

[210]   F. Pereira and T. Alpert. MPEG-4 video subjective test procedures and results. *IEEE Transactions on Circuits and Systems for Video Technology*, 7(1):32–51, February 1997.

Index